MW01517674

Advanced Microsystems
for Automotive Applications 2011

Gereon Meyer · Jürgen Valldorf

Advanced Microsystems for Automotive Applications 2011

Smart Systems for Electric, Safe and Networked Mobility

 Springer

Dr. Gereon Meyer
VDI/VDE Innovation + Technik GmbH
Steinplatz 1
10623 Berlin
Germany
gereon.meyer@vdivde-it.de

Dr. Jürgen Valldorf
VDI/VDE Innovation + Technik GmbH
Steinplatz 1
10623 Berlin
Germany
juergen.valldorf@vdivde-it.de

ISBN 978-3-642-21380-9   e-ISBN 978-3-642-21381-6
DOI 10.1007/978-3-642-21381-6
Springer Heidelberg Dordrecht London New York

Coverdesign: deblik, Berlin

Printed on acid-free paper

Springer is part of Springer Science+Business Media (www.springer.com)

Preface

Fundamental transformations are imminent for the automobile: propulsion technologies are going to shift from combustion engines to electric motors; cars will soon be safer than ever before; and traffic will become increasingly efficient on prospectively more intelligent roads. Currently, the most evident future trend is the electrification of the car, which means the replacement of two major building blocks of its propulsion system, the engine and the gas tank, by completely different technologies, namely an electric motor and a battery.

One of the unique features of the electric power train is that it is controlled by electronic signals rather than by mechanical forces. The flow of energy and information between batteries, motors and wheels, and at the interface to the power grid can thus be managed by smart systems. So far, distributed functionalities can be easily integrated into one single subsystem, for example the intelligent wheel. This helps to optimize the energy efficiency and the driving range of the electric vehicle. Smart systems can successfully address challenges arising from the increased level of integration such as safe operation of a wheel carrying out acceleration, breaking and energy recovery functionalities at the same time.

It is often said that electrification requires more information and communication technologies to be integrated into the car. However, it can also be seen as an opportunity to fundamentally review the electric and electronic architecture of the automobile, and to generally reduce the complexity of propulsion, safety and comfort functions. In combination with the full potential of networking capabilities, this may lead to a completely new platform for sustainable and connected individual mobility.

The papers published in this book cover novel components, future architectures and smart systems that enable the automobile and road transport of the future. They have been selected from the submissions to the 15th International Forum on Advanced Microsystems for Automotive Applications (AMAA 2011) "Smart Systems for Electric, Safe and Networked Mobility" held in Berlin (Germany) on 29 and 30 June 2011. Organizers of the AMAA are VDI/VDE Innovation + Technik GmbH on behalf of the European Technology Platform on Smart Systems Integration (EPoSS), supported by the two EU-funded Coordination Actions of the Public Private Partnership European Green Cars Initiative, ICT4FEV and CAPIRE.

We would like to thank all AMAA authors for their time and effort to summarize the excellent results of their recent work and to present these to the worldwide community of automotive engineers and managers from the industry as well as academic scholars at the conference. We are also much indebted to the members of the AMAA Steering Committee for their important help in selecting the best papers and speakers. And, finally, we would like to acknowledge the substantial support provided by industrial sponsors and public funding authorities, particularly the European Commission.

In our role as editors and conference chairs of the AMAA 2011 we are very thankful for the great assistance provided by an enthusiastic team of colleagues at VDI/VDE-IT. In particular, we would like to thank Iohanna Gonzalez for her engagement and commitment in running the AMAA office, and Rene Stein and Anita Theel for all the hard and excellent work of preparing the book at hand. Last but now least, we would like to express our great appreciation to Wolfgang Gessner for his most valuable advice and continuous support.

Berlin, June 2011

Dr. Gereon Meyer
Dr. Jürgen Valldorf

Funding Authority

European Commission

Supporting Organisations

European Council for Automotive R&D (EUCAR)

European Association of Automotive Suppliers (CLEPA)

Advanced Driver Assistance Systems in Europe (ADASE)

Zentralverband Elektrotechnik- und Elektronikindustrie e.V. (ZVEI)

Mikrosystemtechnik Baden-Württemberg e.V.

Hanser Automotive

enabling MNT

Organisers

European Technology Platform on Smart Systems Integration (EPoSS)

Coordination Action "Information and Communication Technologies for the Full Electric Vehicle"

Coordination Action "PPP Implementation for Road Transport Electrification"

VDI|VDE Innovation + Technik GmbH

Honorary Committee

Eugenio Razelli

President and CEO,
Magneti Marelli S.P.A., Italy

Rémi Kaiser

Director Technology and Quality
Delphi Automotive Systems Europe, France

Nevio di Giusto

President and CEO
Fiat Research Center, Italy

Karl-Thomas Neumann

Executive Vice President E-Traction
Volkswagen Group, Germany

Steering Committee

Mike Babala	TRW Automotive, Livonia MI, USA
Serge Boverie	Continental AG, Toulouse, France
Geoff Callow	Technical & Engineering Consulting, London, UK
Bernhard Fuchsbauer	Audi AG, Ingolstadt, Germany
Kay Fürstenberg	Sick AG, Hamburg, Germany
Wolfgang Gessner	VDI|VDE-IT, Berlin, Germany
Roger Grace	Roger Grace Associates, Naples FL, USA
Klaus Gresser	BMW Forschung und Technik GmbH, Munich, Germany
Horst Kornemann	Continental AG, Frankfurt am Main, Germany
Hannu Laatikainen	VTI Technologies Oy, Vantaa, Finland
Günter Lugert	Siemens AG, Munich, Germany
Roland Müller-Fiedler	Robert Bosch GmbH, Stuttgart, Germany
Paul Mulvanny	QinetiQ Ltd., Farnborough, UK
Andy Noble	Ricardo Consulting Engineers Ltd., Shoreham-by-Sea, UK
Pietro Perlo	Fiat Research Center, Orbassano, Italy
Detlef E. Ricken	Delphi Delco Electronics Europe GmbH, Rüsselsheim, Germany
Christian Rousseau	Renault SA, Guyancourt, France
Patric Salomon	4M2C, Berlin, Germany
Florian Solzbacher	University of Utah, Salt Lake City UT, USA
Egon Vetter	Ceramet Technologies Ltd., Melbourne, Australia
David Ward	MIRA Ltd., Nuneaton, UK
Hans-Christian von der Wense	Freescale GmbH, Munich, Germany

Conference Chairs:

Gereon Meyer	VDI|VDE-IT, Berlin, Germany
Jürgen Valldorf	VDI|VDE-IT, Berlin, Germany

Table of Contents

Electrified Vehicles

Safety & Driver Assistance

Networked Vehicle

Components & Systems

Electrified Vehicles

Partial Networking in the Electrical Vehicle

S. Müller, B. Elend, NXP Semiconductors Germany GmbH

Abstract

When talking about electrical mobility, we mainly think of extending the cruising range of vehicles. Energy management works while driving, charging, or parking. System requirements for e.g. energy saving brings management of control networks to a higher complexity level and results in new and extended requirements for semiconductor devices. These are mainly "selective wake-up capability" to realize Partial Networking (PN) and "longer product lifetime", e.g. battery charge cycle adds to time of active drive. A PN standard for high-speed CAN physical layer is developed by the SWITCH group, a composition of car makers and semiconductor suppliers and is planned as extension to ISO11898. In networks, featuring PN, electronic control units wake up from sleep mode when a certain wake-up message is detected. Compared to existing ISO 11898-5 conform CAN physical layer products, additional functionality in the transceiver is needed to detect wake-up commands. This has to operate properly in a harsh environment of electrical vehicles (EV).

1   Introduction

What is important for EVs? What are trends in in-vehicle networking when we compare an EV to a conventional car with combustion engine? To what extend can paradigms be sustained that have been built with conventional cars?

The following summarizes the system challenges of an EV with impact on in-vehicle networking:
- ▶ Mobility: predictable cruising range, telematics, energy efficiency, size, and weight
- ▶ Lifetime and safety: introduction of new safety-relevant embedded systems
- ▶ System complexity: new energy sources and new power train result in new network demands
- ▶ Robustness: harsh automotive environment, fast transients in power electronics in electrical drive

▶ Isolation towards human interface: high voltages above 60V DC across the vehicle network

2 Networking Trends within Electrical Vehicles

2.1 Network Domain Boundaries are Newly Set

Safety requirements and power saving in EVs are main drivers for in-vehicle networking architectures. Traditional categories in conventional cars are body, chassis, and powertrain. The EV system partitioning is mainly based on the used voltage levels. Background is the need of isolation and preparation to handle high voltage in safety-critical situations such as an accident. Of course, the traditional categories keep their relevance. Here, we see the trend that more and more functions are distributed amongst electrical control units, e.g. telematics, driver assistance.

2.2 Networks have Longer Duty Time

The EV never sleeps. Its network is never completely switched off while networks of conventional cars shut down when parking. EVs have, again, to be alert for critical situations such as failures in the system or the high-voltage battery or car crash; hits another car or is hit. Depending on the incident, the system is prepared to prevent battery from deep drainage and then damage or to separate high-voltage battery from rest of the vehicle for safety reasons.

The anticipated role of an EV as a medium in the public energy grid to store energy [1] illustrates as well that an EV will have extended operation with the energy network and accordingly communicates with the "grid" while parking.

2.3 Network Management Implementation Principle is Changed

Network management in conventional cars expects modules to distribute status messages on a regular base. Does this comply with the need for energy efficiency? We anticipate a move to an "event-trigger based" implementation, i.e. showing activity in the network only when needed. Number of maintenance messages is reduced to a minimum.

2.4 Network is Part of the Energy Management System

Some EV functions are always in operation (e.g. battery monitoring, energy management) and create bus traffic. This keeps modules in CAN networks active even when these modules do not contribute to a.m. system functions. A mechanism is needed that allows switching off/on functions while other functions remain active and exchange data via network, optimized for operation modes such as drive, park, and charge. Fatal for the energy balance if all modules wake up by bus traffic. This impacts the energy balance in a negative way and reduces cruising range [2]. PN provides the necessary feature for networks to switch off modules and quickly reactivate these when needed.

For example, a journey with the EV is scheduled and energy is to be stored in upfront in the battery. Before charging the battery, the amount of energy will be calculated that is needed to drive to the wished location. Navigation and traffic information functions are switched on for the route calculation. When route calculation is finished, results are communicated via the communication network and then navigation and traffic functions are switched off. Charging of the battery starts and accordingly only the involved functions for this operation communicate.

2.5 Paradigm Change in Networking from Conventional Car to EV

▶ Safety aspects dominate architecture and network choice (separation of voltage domains)
▶ Control network becomes an important means of energy management in the vehicle
▶ Parts of the control network are always active

3 What is Partial Networking About?

The usage of PN in conventional cars is typically with comfort modules (functions) that can be switched off during drive or are to be configured during start of the car. Some functions are still expected to be available when ignition of the conventional car is turned off. Examples are trunk lift, seat, window lifter, pre- or auxiliary heating, and sunroof. As described before, we expect a paradigm change within EVs. PN will become an important part of the energy management system. Easiness of implementation, robustness, and attached costs are criteria to successfully employ the PN feature in EV architectures on HW and SW (module) as well on system (network) level.

3.1 Definition

The ability to operate a certain part of a network in a certain moment is called PN. See Fig.1 where a green box means a module is switched on and a gray box that a module is switched off (right car). In ISO11898-5 CAN networks, all modules are switched on when at least two modules communicate (left car). Exceptions are realized by today by switching off supply of a selected module or by using dedicated wake-up wires. Each option is hard-wired and does not offer flexibility in its configuration. With PN, modules wake up by a certain message sent via network.

3.2 Standardization of Partial Networking

German car makers initiated the SWITCH (Selective Wake-able and InteroT†or vendors like NXP joined this interest group. SWITCH developed between July and December 2010 a draft for the extension of ISO11898 introducing a new wake-up mechanism. In short, a valid wake-up message is detected when the received ID matches to a predefined ID, the received data length code matches to the predefined data length code, and the received data field matches to a predefined data field content.

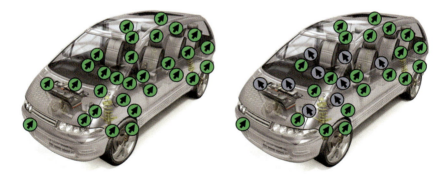

Fig. 1.   Vehicle without and with Partial Networking

3.3 Partial Networking Transceiver Architecture

From the system perspective, modifications in HW (control mechanism) and SW (network management extension) are needed to implement the PN feature. We start with the HW architecture changes of a transceiver.

In order to realize the selective wake-up function, the receiving part of a CAN protocol controller has to be integrated into a PN transceiver as well as the oscillator that clocks this internal protocol controller. Compatibility to standard transceivers in SO14 package is required by the German car makers. Therefore, there is no option to connect an external oscillator like crystal or ceramic resonator to the transceiver. Such external component requires more supply current than an integrated oscillator and would be in conflict with power saving targets. Furthermore, would add costs and space to a printed circuit board. Fig.2 depicts exemplarily the new transceiver architecture.

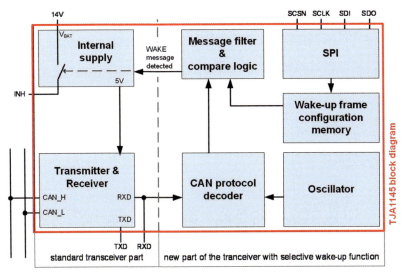

Fig. 2. Exemplary Transceiver Architecture TJA1045 for Partial Networking

In fact, all functional blocks except the transmitter need to be supplied directly from battery because they need to be operational also in low power modes (standby and sleep) when 5V supply is switched off in the according module.

If activity occurs on the network that wakes up ISO 11898-5 conform transceivers, the PN transceiver will not signal a wake event on RxD and INH pins. However, would activate receiver, protocol decoder, oscillator, and message filter & compare logic. If the bus remained silent for a certain period in time, these blocks would get deactivated, again. The wake-up event is signaled on RxD and INH in case the configured wake-up message has been received.

Overall, the challenge for the hardware implementation of PN is to find an on-chip oscillator design of certain accuracy, i.e. with perfect compensation for temperature, supply voltage variation, production spread, and ageing in order to comply with the robustness requirements of the harsh environment of an EV.

3.4   Network Management Modifications

Besides HW modifications, the PN implementation requires changes in the network management. This impacts different levels of the SW architecture. An instance such as a gateway needs to keep track which modules have been switched off on purpose or due to error condition. These questions are addressed in subgroup "Efficient Energy Management" of Autosar work package WP-1.1.1. PN functionality is available with Autosar release 3.2.1 [3].

Fig.3 summarizes SW elements implemented in Autosar standard transceiver driver level (left hand) and needed additions to support PN (right hand). These are APIs, SPI support package, wake-up reasons, a PN configuration container for general PN support, wake-up frame configuration (ID, DLC, mask, data etc.), and baud rate. Important is the support of different shutdown sequences for PN transceivers, as the bus is not idle during network shutdown.

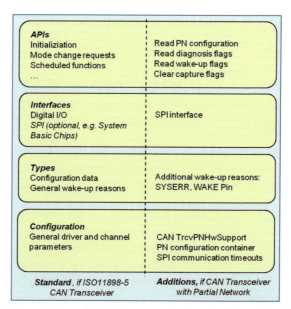

Fig. 3.   Autosar Release 3.2.1, Incorporation of CAN Transceiver Driver

3.5 Architectural Changes on Module Level

The CAN network architecture in the vehicle as well as the HW architecture on module level does not change when PN is introduced. The PN transceiver with its selective wake-up function is responsible for detecting the wake-up event on the network and controls the activation of the voltage regulators for the entire module. This is identical to the operation of a standard transceiver according to ISO11898-5. Fig.4 shows how a standard High-speed CAN transceiver like the TJA1041 or TJA1043 can easily be replaced on module level by a PN transceiver TJA1145. However, since the configuration of the wake-up message is necessary, the TJA1145 features a SPI interface instead of having error (ERRN) and mode control pins (STBN, EN).

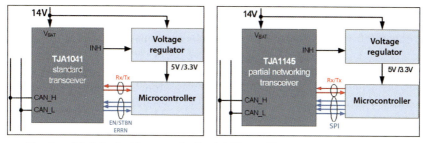

Fig. 4.   Module Architecture for Partial Networking

3.6 Benefit of the Introduction of Partial Networking

PN does not require a new network or module HW architecture, will be standardized in ISO11898 as well as in Autosar, offers new functions that are used in conventional cars to increase comfort but also to comply with new governmental rules for energy saving. These advantages can be used in EVs for the implementation of a robust energy management system and ends up in an increased cruising range.

4   Relevance of Partial Networking for Electrical Vehicles

4.1   Mobility

Today, we do not know where we will end up with the cruising range of EVs. Interesting fact is that an EV already in 1909 could drive 259 km with one charge, 1911 already 324 km, and the Tesla Roadster in 2009 raised the bar with more than 500 km [4]. However, experts expect within the next decade steep improvements in the power density of batteries. Needless to say that each "saved" Watt, directly contributes to the cruising range of an EV. PN excellently contributes with a robust and reliable approach to the energy balance of an EV. Industry expects that power savings in a conventional car may sum up to 70 Watts [5] and is a first reference for EVs. What this means for the extended cruising range depends on the EV characteristics and the efficiency of the chosen EV architecture.

4.2 Safety and Lifetime

EV stands for the introduction of safety-relevant embedded systems, not reusable from the conventional car. A typical EV has three main modes: drive, charge, and park. While in "park" with conventional cars, all functions are switched off, the EV keeps safety-relevant functions active. This links to the battery stack that is always on and is forced to zero leakage in case of a failure in order to avoid damages of the battery by deep discharge. Another aspect to keep the safety sub-system always active is to detach the high-voltage battery from the rest of the vehicle in case of a car crash. Tab.1 shows main system modes versus sub-systems then in operation. As a consequence, the safety sub-system needs to follow extended product lifetime tests.

Mode	EV Sub-System [x=On]		
	Engine	Charger	Safety
Drive	✗		✗
Charge		✗	✗
Park			✗

Tab. 1.   System Modes and Sub-systems in Operation

4.3 Complexity and Lifetime

A discussion took place in the beginning of the SWITCH group how to implement the PN mechanism. On the short list were two options to detect the wake-up message: 1. Detection by CAN controller that is kept active while the rest of the microcontroller, in which it is nested, stops or 2. Add a reduced protocol engine to the silicon of the transceiver. The expert community of the car makers voted for the second option and limited with this changes in the entire system. With this, we anticipate that on device level the above mentioned product lifetime extension is applied to only one device, the transceiver, but not to the microcontroller, voltage regulators, capacitors, etc. While the requirements on lifetime have not been concluded, yet, a first indication from car makers is around to triple the lifetime testing. This will add to the product development lifecycle as well as to the final product cost for the device. However, and again, good if we limit the number of impacted devices, the PN draft standard does.

4.4 Robustness

When we talk about robustness of in-vehicle networking, we basically think about EMC performance in terms of immunity. The good news is that in the last years big steps have been made and the gained knowledge can also be applied to PN transceivers. Moreover, the SWITCH group has already defined dedicated PN EMC requirements. The bad news is that some experts see in EVs an increasing challenge for EMC due to high voltage and high current transients leading to hazardous electromagnetic fields not known from conventional cars. Thus ISO7637 impulse immunity during operation might become one of the new challenges for the semiconductor suppliers. What does robustness of PN in EVs mean? Do wake-up messages have different vulnerability than other messages? Yes, reason is the fact of two separate reception paths. 1. Potential wake-up messages are received and decoded by transceiver with on-chip oscillator. Power consumption in this reception path is limited to a very low value. Directly connected to battery, supply lines might be disturbed by transients. 2. In normal operation, messages are decoded by microcontroller that is connected to quartz as a reliable clock source. The receiver in the transceiver consumes more power and suppresses noise and stabilizes the supply. It is too early to compare robustness of "wake-up message detection in a PN transceiver" in all PN implementation concepts but already clear that the critical factor is the on-chip oscillator and its resulting stability despite distortions like electromagnetic fields, ringing, sender clock tolerances, as well as cranking pulses on the supply. With the TJA1045, NXP found a smart implementation.

5   Summary

▶ EV enables new dimension of efficient driving and need for extended energy management

▶ Lifetime aspects in EVs are tremendously important due to embedded safety systems

▶ Paradigm change in networking: conventional car to EV is from comfort to energy management

▶ Partial Networking is excellent means for energy management; parts of EV are always active

▶ Partial Networking contributes into all operation modes of EVs: drive, charge, and park

▶ Multiple disciplines are involved in Partial Networking and standards are driven for HW and SW

▶ Robustness will be key differentiator between PN transceivers; accuracy of the on-chip oscillator

References

[1]   Sauer, D. U., Auslegung von Elektrofahrzeugen und Energiemanagement im Netzverbund, ZVEI Kompetenztreffen Elektromobilität, p.24 ff, 2009.

[2]   Spannhake, S., Wirkungsgradoptimierung von Elektrofahrzeugen auf Gesamt-systemebene, VDI Fachkonferenz Elektromobilität, p.17, 2010.

[3]   Bunzel, S., Autosar Release News, October 2010.

[4]   Staretz, D., Von der Historie zur Hysterie, Das Elektroauto hatte schon vor hundert Jahren seinen Welterfolg, Profil.at, Volume 50, p.80, 2010.

[5]   Elektronik im Kraftfahrzeug, VDI-Berichte 2075, ISBN 978-3-18-092075-7, p.127, 2009.

Steffen Müller, Bernd Elend
NXP Semiconductors Germany GmbH
Stresemannallee 101
Hamburg
Germany
st.mueller@nxp.com
bernd.elend@nxp.com

Keywords: electrified vehicles, components and systems, partial networking, SWITCH, autosar, transceiver level driver, selective wake-up, transceiver architecture, ISO 11898

Development of Mathematical Models for an Electric Vehicle With 4 In-Wheel Electric Motors

E. Cañibano Álvarez, M. I. González Hernández, L. de Prada Martín, J. Romo García, J. Gutiérrez Diez, J.C. Merino Senovilla, Fundación CIDAUT

Abstract

CIDAUT Foundation is a Spanish non-profit Research and Development Centre for Transport and Energy. One of CIDAUT´s current lines of work is sustainable mobility, involving the electric vehicle and its infrastructure. In order to advance in this direction, CIDAUT has designed and manufactured a technological demonstrator consisting of an Electric Vehicle. Currently, CIDAUT's research deals with the implementation of Direct Yaw Moment Controls (DYC) in electric vehicles with 4 in-wheel motors. This paper aims to describe the different simplifications and subsystems created to generate advanced control algorithms. This technology belongs to the electronic active systems equipment incorporated in the future in order to have a proper control of the behaviour of the vehicle. This new powertrain system presented in this paper allows control engineers to develop more advanced algorithms due to highly versatile systems where is possible to generate forces independently on each wheel.

1   Introduction

Nowadays, the high price of the electronic sensors makes engineers build active safety systems based on the knowledge of a reduced number of real vehicle states. As a consequence, many other ones have to be estimated based on simplified mathematical models that manage to reproduce reality. The state of the art presents several different approaches and final targets. In this case, the new possibilities that an independently controlled wheel traction system offer are explored. There exist three remarkable advantages of in-wheel traction systems: motor torque generation is quick and accurate, motors are easily installed in two or four wheels and motor torque can be known precisely. These advantages enable us to easily implement antilock braking and traction control systems, chassis motion control like Direct Yaw Control (DYC) and an estimation of road surface condition.

Hence, yaw moment reference signal detection can be followed not only by braking, but also by giving an increasing torque to wheels in order not to lose any global vehicle velocity.

2 Experimental Vehicle and Numerical Model

A conventional internal combustion engine (ICE) vehicle was used as starting point to develop the electric vehicle technological demonstrator. The first relevant characteristic of this vehicle is the use of two independent motors in the rear axle (Fig. 1.). Each motor is governed by one controller and both are connected to the PLC (Programmable Logic Controller) of the vehicle.

The PLC of the vehicle receives the signals concerning the motors speed as well as the steering wheel position. With this information the PLC calculates the signal to be sent to each motor. Depending on the parameters introduced by the user, the dynamic behaviour of the vehicle can be understeering, oversteering or neutral. The PLC also allows the user to select the kind of driving: snow, economic or dynamic.

Fig. 1.   Experimental demonstrator developed by CIDAUT

Fig. 2.   MSC. ADAMS/Car model with in-wheel electric motors

The numerical model generated in MSC.ADAMS/Car MDR3 has been built as similar as possible as the real car. Every subsystem has been created in order to have the same dynamic behaviour and they allow us to make correlations of the experimental results. The powertrain subsystem has been built to reproduce the characteristics of the in-wheel motors. Besides, the control algorithm is developed in Matlab/Simulink R2010a to imitate the real components behaviour.

3   DYC System: Logic Description and Components

The motivation for the development of yaw control systems comes from the fact that the behaviour of the vehicle at the limits of adhesion (rain, low μ road, etc.) is quite different from its nominal behaviour. This kind of system pretends to reduce the deviation of the vehicle behaviour from its nominal behaviour and also prevent the vehicle slip angle from becoming large.

The main goal of the control system is to follow the reference yaw moment dictated by a simplified two d.o.f model. As shown in the following figure, the control model is compound of different modules that can be grouped as: direct calculations, estimated states and logic modules. Each of them is described in the next lines to build a complete and easy to follow and tune control system.

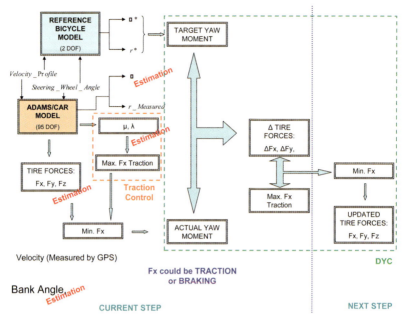

Fig. 3. Global DYC control algorithm scheme

3.1 Reference Model

The 2 d.o.f bicycle model is mainly selected by researchers as the reference model to calculate the yaw rate and side slip angle signals. Starting from measured values of longitudinal velocity and steering wheel angle, yaw rate and side slip angle reference signals are calculated in order to be followed as close as possible by the controlled system.

$$\begin{bmatrix} \dot{\beta} \\ \dot{r} \end{bmatrix} = \begin{bmatrix} -\dfrac{C_{\alpha F} + C_{\alpha R}}{m \cdot v_x} & -\dfrac{a \cdot C_{\alpha F} + b \cdot C_{\alpha R}}{m \cdot v_x^2} - 1 \\ -\dfrac{a \cdot C_{\alpha F} - b \cdot C_{\alpha R}}{I_z} & -\dfrac{a^2 \cdot C_{\alpha F} + b^2 \cdot C_{\alpha R}}{I_z \cdot v_x} \end{bmatrix} \cdot \begin{bmatrix} \beta \\ r \end{bmatrix} + \begin{bmatrix} \dfrac{1}{m} \cdot C_{\alpha F} & \dfrac{1}{m} \cdot C_{\alpha R} \\ \dfrac{1}{I_z} \cdot a \cdot C_{\alpha F} & -\dfrac{1}{I_z} \cdot b \cdot C_{\alpha R} \end{bmatrix} \cdot \begin{bmatrix} \delta \\ 0 \end{bmatrix} \tag{1}$$

Once the yaw rate is obtained, it is numerically derived and multiplied by the yaw inertia of the vehicle to finally obtain the reference yaw moment. This variable defines the behaviour of the vehicle during the manoeuvre of cornering.

3.2 Estimation of States

Several real vehicle states need to be estimated so as to measure in vehicle dynamics terminology what the vehicle is actually doing. Standard medium class vehicles only incorporate sensors for measuring wheel speed, throttle signal, yaw rate, brake pressure, longitudinal and lateral accelerations and steering wheel angle. As a result, numerous other variables have to be calculated in an easy manner to be integrated in the control algorithm. In the next lines, these simple models are described briefly just to show how many tools are needed.

▶ Estimation of the side slip angle of the vehicle at its cog: Taking the basic equations of the transient bicycle model, the lateral acceleration has two components. As a result, the integration of the variation of lateral velocity allows calculating an approximation of the sideslip angle. This method obtains accurate results for simulation data. For real data, sensor errors and noise must be considered so as to introduce filtering or other data treatment techniques.

$$a_y = \dot{V}_y + r \cdot V_x \quad V_y = \int (a_y - r \cdot V_x) dt \qquad \hat{\beta} = \arctan(V_y / V_x) \qquad (2)$$

▶ Normal tire forces: Steady-State weight transfer models (lateral and longitudinal) have been used to determine the normal forces on each of the four tires. The next step consists on assigning over each tire the corresponding load, according to the reference system selected.

▶ Slip angle of each tire: Having the values of longitudinal and lateral velocities, yaw rate and other geometrical parameters of the vehicle, it is possible to accurately estimate the slip angle on each tire based on simple trigonometric formulas [1]. This calculations are obtained based on the bicycle model.

▶ Lateral and longitudinal tire forces: a Magic Formula Monte Carlo version tire model is used. The coefficients are fitted according to convenient test carried out.

▶ Electric motor model and longitudinal force transmission estimation: The conversion of the driver throttle signal into traction torque on each wheel is an important issue for this model. The maximum transmissible force between tire and road is compared to the maximum force that could give the motor in this situation. Once the longitudinal slip is calculated, the value of the friction coefficient is obtained through a lookup table (See Figure 4). This is a simplification of the real model but is also widely used in vehicle dynamics control algorithms [5]. Having the values of normal and lateral force for each time step, using the friction ellipse approximation is possible to calculate the maximum transmissible longitudinal force to the road.

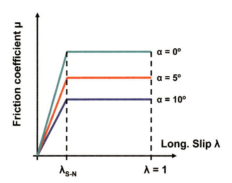

Fig. 4. Ideal (simplified) curve for friction coefficient versus longitudinal slip and slip angle

▶ In parallel, the torque and, therefore, the longitudinal force offered by the electric motor at the current conditions are intended. Making use of the torque vs. angular velocity map of the motor, the measured velocity of each of the wheels allows obtaining the maximum torque in this situation. Then this torque is multiplied by the throttle signal to obtain the final value really developed.
▶ The last step is to compare the both values previously obtained. The real value developed between the road and the tire is the minimum of them.
▶ GPS/INS systems are seen to be a near future benchmark for road vehicles. For this reason, their information could be used to increase the accuracy of the control systems. They provide information about the position and velocity of the car which could be really helpful. This improvement will be included in future version of the control system.

3.3 ABS and TCS Systems

Antilock Braking (ABS) and Traction Control (TCS) systems are becoming standard systems for advanced vehicle control systems. Both systems are implemented in such a way that they do not work under a velocity threshold imposed by the programmer.

The ABS follows a logic that only allows to work with incremental pressures (relative to the value of the previous instant). This statement means that there exist only three possible actuations as response to the measurements: increase, decrease or maintain the pressure in the hydraulic braking system. Therefore, it is needed to calculate the increment of brake torque to add in each time step. This value will be a function of the angular acceleration and it is possible to differentiate between three ranges:

▶ $b\_abs < \alpha < a\_abs$: The brake pressure defined by driver and ESP is maintained.

▶ $\alpha \le a\_abs$: The wheel is close to get blocked so the brake torque must be reduced.

▶ $\alpha \ge b\_abs$: The wheel has a pretty high angular acceleration and it is needed to increase the brake torque.

As can be deduced from the previous comments, in the range of wheel angular acceleration where the ABS does not actuate, the final pressure matches with the required value form the driver and the ESP. On the other hand, if the values a_abs or b_abs are exceeded, a negative or positive increment is applied respectively to the brake pressure. Finally, the brake pressure is limited by the physical boundaries of the brake system. Being an electrical device, the in-wheel motors do not follow this actuation logic. The sample frequency of the system is usually around 100 Hz.

3.4 DYC Algorithm

The distribution and type of torque is firstly done based on whether the car is in an acceleration or a braking manoeuvre. According to this state, the yaw moment generated will be such that the vehicle control is lost. The increment of yaw moment is created only from longitudinal forces because an active steering system has not yet been developed.

The decision of using traction torque (instead of braking) is made based on the value of the throttle signal in each instant. The bound value is adjusted not to have velocity loses, as usually happens in current systems based on braking torques exclusively.It is obvious that the necessary yaw moment could exceed the achievable one generated by the tires. In this case the system is saturated by the limit of longitudinal forces that is able to transmit to the road.

The following table resumes the actuation logic on each wheel and depending on the working system. In case of understeering or oversteering (controlled by the difference of the nominal and current yaw moment in the vehicle), each independently controlled wheel will be accelerated of braked to follow the reference given.

	UNDERSTEERING	*OVERSTEERING*
BRAKING SYSTEM	• Brake front inner wheel	• Brake front outer wheel
	• Accelerate rear outer wheel (similar to decrease brake torque)	• Accelerate rear inner wheel (similar to decrease brake torque)
TRACTION SYSTEM	• Accelerate front outer wheel	• Accelerate front inner wheel
	• Brake rear inner wheel (similar to decrease traction torque)	• Brake rear outer wheel (similar to decrease traction torque)

The conversion of yaw moment into traction or braking torque is obtained from a gain value calculated from the simple equilibrium of moments of the vehicle in a straight line.

To simulate the response time of the actuators, delays are added to the response action of the subsystems. The time constant introduced consists of a typical reaction time for hydraulic systems. For electric equipment, the estimated reaction time is assumed to be the half of the previously commented.

4   Conclusion

As a result from M2IA project, CIDAUT has developed a technological demonstrator and is researching new control system possibilities. There exist wide possibilities to develop as each wheel is able to be controlled independently. The benefits that active safety systems can obtain from them are still to be exploited. Once electric vehicles are widely spread, vehicle manufacturers will opt to optimize the vehicle handling. This consists on an initial approach to the future advanced safety system algorithms. Simple models are presented in this paper to model an advanced direct yaw moment control system to be implemented in a vehicle with in-wheel motors.

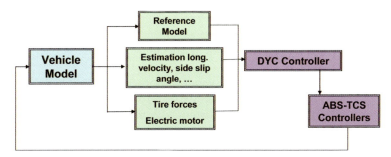

Fig. 5.   SIMULINK diagram of the DYC control algorith

References

[1] Esmailzadeh, E. , Vossoughi, G. R. and Goodarzi, A. "Dynamic modelling and analysis of a four motorized wheels electric vehicle", Vehicle System Dynamics, Vol. 35, No. 3, pp. 163-194, 2001.

[2] Hori, Y., "Future Vehicle Driven by Electricity and Control- Research on Four-Wheel-Motored 'UOT Electric March II'", IEEE Trans. Ind. Electronics, Vol. 51, 954-962, 2004.

[3] Kiencke U., Nielsen L., "Automotive control systems for engine, driveline and vehicle", Springer-Verlag, 2nd Edition, Berlin, 2005.

[4] Rajamani, R. "Vehicle dynamics and control", Springer-Verlag, 2006.

[5] Bosch, R., "Safety, comfort and convenience Systems", John Wiley & Sons, Plochingen, 2006.

[6] Ehsani M., Gao Y., Gay S. E., Emadi A., "Modern Electric, Hybrid Electric, and Fuel Cell Vehicles: Fundamentals, Theory, and Design", CRC Press, Boca Raton, FL, 2005.

[7] Larminie J., Lowry J., "Electric Vehicle Technology Explained", John Wiley & Sons, Ltd., West Sussex, 2003.

[8] Guillespie T. D., "Fundamentals of Vehicle Dynamics", Society of Automotive Engineers, Inc., Warrendale, PA, 1992.

E. Cañibano Álvarez, M. I. González Hernández, L. de Prada Martín, J. Romo García, J. Gutiérrez Diez, J.C. Merino Senovilla
Fundación CIDAUT
Parque Tecnológico de Boecillo P209
47151 Boecillo
Spain
estcan@cidaut.es
mangon@cidaut.es
luipra@cidaut.es
javrom@cidaut.es
javgut@cidaut.es
juamer@cidaut.es.

Keywords: in-wheel motor, electric vehicle, control algorithm, MSC.ADAMS/Car, Mechatronics, Simulink

Monitoring and Testing of an Electric Vehicle With 2 In-Wheel Motors: Variable Dynamic Behaviour and Control

E. Cañibano Álvarez, M. I. González Hernández, L. de Prada Martín, J. Romo García, J. Gutiérrez Diez, J.C. Merino Senovilla, Fundación CIDAUT

Abstract

CIDAUT Foundation is a Spanish non-profit Research and Development Centre for Transport and Energy. One of CIDAUT´s current lines of work is sustainable mobility, involving the electric vehicle and its infrastructure. In order to advance in this direction CIDAUT has designed and manufactured a technological demonstrator consisting of an electric vehicle. This vehicle is been used to test different power train configurations, batteries technologies, electronics, and specially dynamics behaviour and control algorithms. The constructive solution chosen is 2 rear in-wheel motors as one of the most versatile configurations. CIDAUT has developed mathematical models to reproduce the dynamic behaviour of such vehicles. These models allow full control of the dynamic behaviour while simulating different control algorithms in order to improve active safety and maintain "fun to drive" (based on the control parameters of the in-wheel motors). CIDAUT's technological demonstrator has been used to validate those models and control algorithms. The three basic configurations used in the experiments are no control, low speed control and high speed control.

1 Introduction

As part of its effort towards Sustainable Mobility, CIDAUT Foundation (http://www.cidaut.es) has designed and manufactured a technological demonstrator consisting of an electric vehicle. This vehicle was developed in CIDAUT's research and development project Minimal Environment Impact Mobility (M2IA, Movilidad de Mínimo Impacto Ambiental). The vehicle was developed to be the testing ground for different types of batteries and motor configurations in order to improve powertrain efficiency in electric vehicles.

The next stage for researchers was to develop a numerical model able to accurately reproduce the dynamic behaviour and performance of the vehicle in handling and ride. Several tests are carried out to find the specific behaviour of the electric car. This data is then used to correlate the equivalent simulations.

2   Experimental Vehicle and Numerical Model

Starting from the real vehicle, numerical simulations have been carried out to reproduce its dynamic behaviour. Several test and measures have been necessary to collect all the data required to build a proper model in simulation, such as mass and inertia of components, geometry of suspension and characteristics of the different subsystems. The tire model has been also an issue of study due to the great influence of the tire forces for the performance of the vehicle.

2.1   Experimental Vehicle: Demonstrator

A conventional internal combustion engine (ICE) vehicle was used as starting point to develop the electric vehicle technological demonstrator. The first relevant characteristic of this vehicle is the use of two independent motors in the rear axle (Figure 1). Each motor is governed by one controller and both are connected to the PLC (Programmable Logic Controller) of the vehicle.The PLC of the vehicle receives the signals concerning the motors speed as well as the steering wheel position.

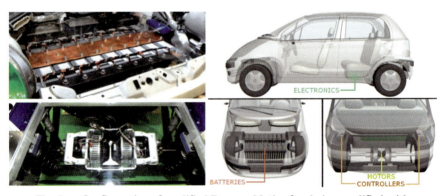

Fig. 1.   Configuration of modified Daewoo Matiz after being modified with two in-wheel electric motors. Real and CAD models

With this information the PLC calculates the signal to be sent to each motor. Depending on the parameters introduced by the user, the dynamic behaviour of the vehicle can be understeering, oversteering or neutral.The PLC also allows the user to select the kind of driving: snow, economic or dynamic.

2.2 Numerical Model in MSC.ADAMS/Car

The model generated in MSC.ADAMS/Car MDR3 has been built as similar as possible as the real car. Every subsystem has been modified in order to have the same dynamic behaviour and it allows making experimental correlations of the results. A new powertrain template has been created in order to reproduce the characteristics of the in-wheel motors. Besides, the control algorithm is developed in Matlab/Simulink R2010a in order to imitate the real components behaviour.Three control algorithms are considered in both real and simulation model vehicles:

▶ No control: Each of the electric motor operates independently according to the throttle signal.
▶ Low Speed Control: According to the Ackerman ratio between the front wheel steered angle, one of the wheels (inner) acts as the master. The slave wheel (outer) develops a different torque related to the previous one to compensate the difference in radius. This way the system achieved a less curved trajectory.
▶ High Speed Control: Basically is an extension of the previous algorithm. The increase in the control signal to the outer trajectory wheel is magnified to achieve a more neutral steer behaviour. This way the understeer performance of the original vehicle is compensated.

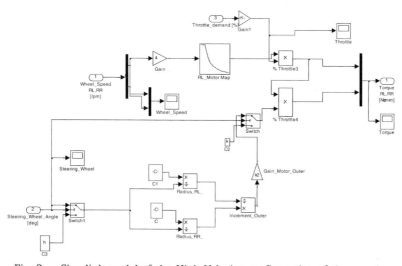

Fig. 2.   Simulink model of the High Velocity configuration of the rear in-wheel motors

The correlation of the model has been made based on the characteristic velocity. This is a specific variable calculated for vehicles which show predominant understeer behaviour. It is defined as the velocity of the vehicle that needs to turn the steering wheel twice to maintain the same radius trajectory. Not only based on the definition graphically, the characteristic velocity is also possible to be calculated from the following formula, where the understeer gradient (K_{us}) appears:

$$V_{characteristic\_US} = \sqrt{\frac{g \cdot wheelbase}{K_{us}}} \tag{1}$$

Being K_{us} the understeer gradient which defines the level of understeer behaviour the vehicle suffers. Its definition comes from the common bicycle model. There exists diverse ways of calculating it. Firstly, based on mathematical relationships could be defined as:

$$K_{us} = \left(\frac{W_F}{C_{\alpha F}} - \frac{W_R}{C_{\alpha R}} \right) \tag{2}$$

The main problem of the use of the previous formula with real test results is the introduction of a single value for the cornering stiffness of the tires. In contrast, it is as well calculated based on the results of a constant radius or a constant steering cornering test. This fact will be developed in the next paragraphs and it does not need to introduce estimated values for any variable.

3   Results

Both experimental test and numerical simulations have been run with the final target of measuring the dynamic behaviour of the vehicle. Moreover, several previously commented control algorithms have been tested just to study the influence of them on the vehicle performance.

3.1   Experimental Test

In order to demonstrate the variations in the vehicle dynamic performance, three different basic configurations were probed. The variations in the dynamic behaviour were controlled by changing the control parameters in the in-wheel motors.

To quantify the results from the different test configurations, a constant radius cornering (CRC) test has been carried out. The radius of the curve is 28 meters. The test consists in increasing the speed of the vehicle through the curve and measuring the evolution of the steering wheel angle. In order to obtain more information also the independent speed of the motors has been registered. The test objective is to determine the influence of electric motor control on handling, as well as to determine if it is possible to improve the vehicle active safety.

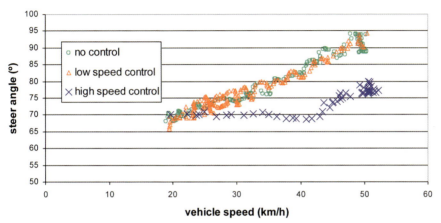

Fig. 3. Representation of the steering wheel angle versus speed for the three configurations in the CRC test

Two main results have been obtained. The first one is the determination of the vehicle dynamic behaviour depending on the control used, and the second one being the maximum speed for the stability loss for each of the control strategies.

It is interesting to note that "no control" and "low speed control" strategies produce almost the same behaviour in the modified vehicle performance. This is due to the fact that the slip of the wheels has not been taken into account for the "low speed" strategy. This influence is important for speeds larger than 20km/h. The possibility of controlling both electric motors independently has allowed the researchers to develop a control law (high speed control) in order to have a neutral behaviour of the modified vehicle. This neutral behaviour can be kept up to 44 km/h, while for higher speeds the modified vehicle behaviour is slightly understeer.According to operating points showed in Figure 3., it can be concluded that the influence of "low speed" control strategy on modified vehicle behaviour is important only for parking maneuvers. For vehicle speeds higher than 20 km/h a more aggressive control strategy is needed.In order to show the results clearly, and bearing in mind that for average values of speed "no control" and "low speed control" strategies behaviour present similar results, only "no control" and "high speed" control strategy results are presented. In Figure 4. the speed of each rear wheel is compared against vehicle speed.

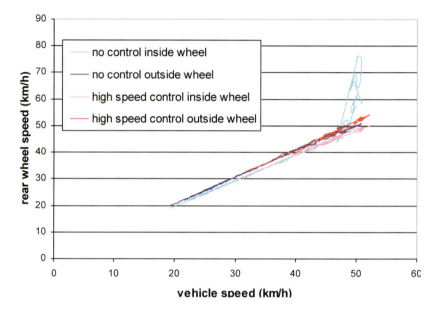

Fig. 4.   Representation of rear wheels speed versus vehicle speed in the
          CRC test

In both cases, for speeds under 48 km/h, it can be seen that exterior wheel speed is higher than inner wheel speed. This difference in the wheel speed values increases when the average speed of the vehicle increases. This behaviour follows that trend up to 54 km/h. In that moment the tires grip limit is reached. No sudden change of direction occurs.

In the "no control" case, when the vehicle speed reaches 48 km/h, the inner rear wheel tends to lose adherence with the road. In this moment there is a longitudinal slip causing the inner wheel to turn faster than the exterior wheel due to the torque from the electric motor and the lack of grip from that wheel to the ground.

3.2 Numerical Simulations

Similarly to the previous real test, the behaviour of the vehicle has been studied from simulation. In this case, as there is not a controller like a driver on the vehicle, simulations are run in open loop mode. This means that a constant radius test can not be carried out because it is not possible to maintain a constant radius trajectory without a controller continuously modifying the steering wheel angle. In real tests, the driver performs the function of the equivalent controller in simulations. In contrast, an equivalent constant steering cornering (CSC) test has been simulated obtaining analogous results for the three configurations in study.

The test is based on an incrementation of the vehicle's speed through the described trajectory and measuring the evolution of the longitudinal velocity and lateral acceleration. Representing the curvature versus lateral acceleration it is possible to obtain the understeer gradient [1].

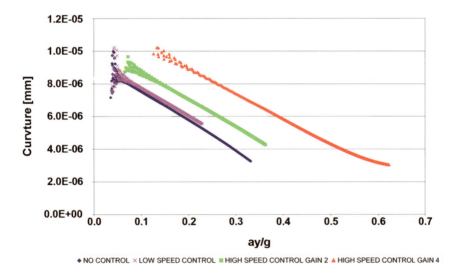

Fig. 5.   Representation of the curvature versus lateral acceleration for the three configurations in the CSC test

The configurations simulated show that neutral behaviour is the direction taken by the control algorithms once the gain of the high speed control increase. These configurations follow the rising characteristic velocity evolution shown in the experimental test. It is also interesting to point out the small difference in behaviour between the configurations of no and low speed control. This fact has been mentioned before as one of the results of the experimental test.

The next step was to reproduce the experimental test to find correlation of the results. For this simulation, it has been used the steering wheel angle from the real test as input data. The radius of the simulated trajectory is pretty similar to the real one with an approximate error of 2 m.

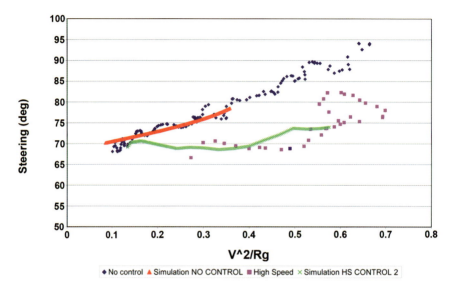

Fig. 6. Representation of the steering wheel angle versus speed for two configurations in the reproduction of the experimental test

Furthermore, it can be seen that, at a similar velocity, the rear inner wheel lifts off the ground and the results are not meaningful from then on. This fact is stated since the normal inner wheel force drops to zero in the same instant.

The graphs show the curves comparison for real and simulated tests carried out in similar conditions and with similar inputs. The reproduction of the results is of high accuracy and also the speed values of the inner wheel contact loss commented in the results of the experimental test.

4   Conclusion

As a result from M2IA project, CIDAUT has developed both a technological demonstrator and a correlated numerical model. It allows control engineers to speed up the process of development of control algorithms. CIDAUT is working in parallel on the development of a Direct Yaw Moment Control system which will use these two vehicles as important tools of progress.

This kind of new electric powertrain devices makes possible to remove the mechanical differential and thus increase the mechanical efficiency and introduce an active stability control. In order to evaluate the efficiency of such powertrain configuration a simple driving test was defined both experimentally and in simulation.

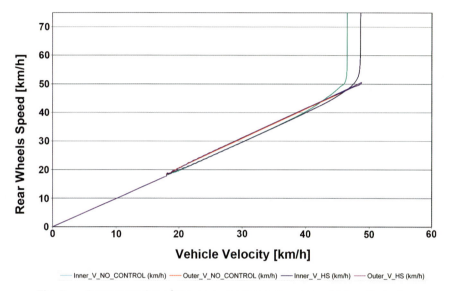

Fig. 7.   Representation of the rear wheel speed versus vehicle velocity for
          two configurations in the reproduction of the experimental test

Some interesting results have been obtained:
  ▶ It is possible to change the behaviour of the vehicle from understeer to
    neutral developing and active control system for the independent elec-
    tric motors.
  ▶ The stability of the vehicle can be improved. The in-wheel motor tech-
    nology opens a wide development field for improving the transmission
    efficiency and the vehicle safety.
  ▶ The versatility of the technology demonstrator allows evaluating the
    efficiency of other powertrain configurations simply by replacing the
    current one and then testing. One of the projects actually under evalua-
    tion is the construction of a demonstrator with four in-wheel motors.
  ▶ The numerical model developed helps to the design of new ideas and
    save costs of development. In addition, reduce the progress time due to
    the fact that it allows easily running many configurations.

References

[1]   Wong, J.Y., 'Theory of Ground Vehicles' 3rd Ed., John Wiley & Sons, New York, 2001.

[2]   Esmailzadeh, E. , Vossoughi, G. R. and Goodarzi, A. "Dynamic modelling and analysis of a four motorized wheels electric vehicle", Vehicle System Dynamics, Vol. 35, No. 3, pp. 163-194, 2001.

[3]   Hori, Y., "Future Vehicle Driven by Electricity and Control- Research on Four-Wheel-Motored 'UOT Electric March II'", IEEE Trans. Ind. Electronics, Vol. 51, 954-962, 2004.

[4]   Kiencke U., Nielsen L., "Automotive control systems for engine, driveline and vehicle", Springer-Verlag, 2nd Edition, Berlin, 2005.

[5]   Rajamani, R. "Vehicle dynamics and control", Springer-Verlag, 2006.

[6]   Bosch, R., "Safety, comfort and convenience Systems", John Wiley & Sons, Plochingen, 2006.

[7]   Ehsani M., Gao Y., Gay S. E., Emadi A., "Modern Electric, Hybrid Electric, and Fuel Cell Vehicles: Fundamentals, Theory, and Design", CRC Press, Boca Raton, FL, 2005.

[8]   Larminie J., Lowry J., "Electric Vehicle Technology Explained", John Wiley & Sons, Ltd., West Sussex, 2003.

[9]   Guillespie T. D., "Fundamentals of Vehicle Dynamics", Society of Automotive Engineers, Inc., Warrendale, PA, 1992.

E. Cañibano Álvarez, M. I. González Hernández, L. de Prada Martín, J. Romo García, J. Gutiérrez Diez, J.C. Merino Senovilla
Fundación CIDAUT
Parque Tecnológico de Boecillo P209
47151 Boecillo
Spain
estcan@cidaut.es
mangon@cidaut.es
luipra@cidaut.es
javrom@cidaut.es
javgut@cidaut.es
juamer@cidaut.es.

Keywords: in-wheel motor, electric vehicle, control algorithm, MSC.ADAMS/Car, Mechatronics, Simulink.

Internet of Energy – Connecting Energy Anywhere Anytime

O. Vermesan, L.-C. Blystad, SINTEF
R. Zafalon, A. Moscatelli, STMicroelectronics
K. Kriegel, R. Mock, Siemens
R. John, Infineon Technologies
M. Ottella, P. Perlo, Fiat Research Center

Abstract

The forthcoming Smart Grid is expected to implement a new concept of transmission network which is able to efficiently route the energy produced from both concentrated and distributed plants up to the final user with high security and quality of supply standards. Therefore the Smart Grid is expected to be the implementation of a kind of "internet" in which the energy packets are managed similarly to data packets, across routers and gateways which autonomously can decide the best pathway for the packet to reach its destination with the best integrity levels. In this respect the "Internet of Energy" concept is defined as a network infrastructure based on standard and interoperable communication transceivers, gateways and protocols that allow a real time balance between the local and the global generation and storage capability with the energy demand, also allowing high level of consumer awareness and involvement. This paper presents some basic concept of the Internet of Energy and, in particular, its impact on Electric Mobility.

1 Introduction

The Internet of Energy (IoE) provides an innovative concept for power distribution, energy storage, grid monitoring and communication as presented in Fig. 1. It will allow units of energy to be transferred when and where needed. Power consumption monitoring will be performed on all levels, from locally individual devices up to national and international levels.

The Internet of Energy will therefore provide to the consumers a highly reliable, flexible, resilient, efficient and cost effective power supply network, in particular enabling the full deployment of distributed power sources (i.e. small scaled renewable sources) in combination with large centralized generators.

One of the cutting edge capacities of the Internet of Energy will be the storage of energy for later use which is becoming a necessity for the full utilisation of the capabilities of renewable sources whose output is intrinsically variable and intermittent. One way to achieve this goal is to incorporate electric vehicles (EVs) not only as energy consumers but also as energy providers and storages. The intended end result is a giant power network extending from Norway's hydroelectric reservoirs to solar power plants in southern Europe, an intelligent grid, which uses modern information technology to coordinate energy demand and response with distribution on a real time basis, with the highest possible efficiency and reliability.

The implementation of this will allow energy generation peaks to be stored in the batteries of the connected EVs for later usage, in particular during high power consumption demand periods. A real time monitoring will be essential for both the consumer control of the process on the basis of the real-time energy price, and for the power supplier to restrict or permit to a specified device the access to the power network on the basis of the grid's local and overall state. (As an example during a fast charge, in the timeframe of few minutes an electric vehicle can pull as much power as an entire home consumes during the whole day at full load and this need to be managed, in particular when concurrent activities exist.).

Moreover the electric utilities will need to provide incentives to consumers for shifting loads (the vehicle charging is one of them) during off-peak hours. For implementing this, the consumers will need to have the possibility to program the charger to make the battery be fully charged at the scheduled time of the vehicle mission, still optimising the charging rate in order to get the lowest possible energy price. Smart battery charger will be necessary for the implementation of these features.

Finally the implementation of a smart grid is imperative to allow the market success of electric mobility and to overall reduced costs, increased efficiency, increased consumer awareness, and allowing the network operators to supervise the local and overall load management.

In this respect the smart meter acting, de facto, often as a home area gateway (HAG) will be a key enabler of the specified functionalities, implementing the bridge between the home energy management domain which includes the electric vehicle - across the Local and Wide area Gateways - to the data farms for both user and utility access. Any data can be sent to the utility and the needed information such as charge rate or charging patterns will de delivered using the building Internet connection.

Alternative ways for accessing EV chargers and the connected building/home devices are through smart phones or mobile/tablet PCs using Wi-Fi, ZigBee or NFC/RFID communication protocols.

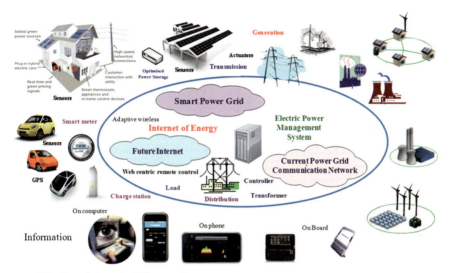

Fig. 1.   Internet of Energy Concept

This paper presents selected advances that are expected in microsystems technologies, power electronics components, communication circuits and embedded systems modules that are to be addressed in the ARTEMIS IoE project whose objective is to develop hardware, software and middleware for seamless, secure connectivity and interoperability achieved by connecting the Internet with the energy grids, and having its first and preferred application in the infrastructure for the electric mobility. The underlying architecture is of distributed Embedded Systems (ESs), combining power electronics, integrated circuits, sensors, processing units, storage technologies, algorithms, and software.

2   Energy Infrastructure

The forthcoming intelligent grid is relying on several innovations which are being implemented in the existing energy infrastructure and in particular the new transmission lines, the communication features, the incorporation of diverse renewable sources, the deployment of distributed storage and storage plants. In the following chapter we will discuss the state of the art and the innovations that are expected in the near future.

2.1 Transmission Line

As per the current state of the art, most electrical energy is transmitted over high voltage AC three phase lines. This poses severe constraints to the maximum distance between the energy producing plant and the first transformation stage, due to the parasitic capacity (dependent from the mutual distance of the wires and from their length) that in turn generates reactive power. Even with large aerial lines, it is impossible to have efficient lines whose length exceeds 1000 km. For underground and underwater lines this maximum distance is much lower.

However, the recent advancements in ICT and in particular in power electronics have made it possible to have, virtually unlimited length, high voltage DC transmission lines. These are currently being implemented in several countries of Europe, and in particular for offshore wind farms [1]. Several projects are being implemented to create a European HVDC super grid [2].

2.2 Sensors

The adoption of electric vehicles, the emergence of dynamic new economies, and the mandated incorporation of alternative power generation put a huge strain on the existing energy infrastructure. All these call for implementing smart grid technologies that range from sensor and actuator technologies to semiconductor, communication and software technologies. The Internet of Energy is characterised by two way digital communication that enhances monitoring and control, allows real time smart metering of consumer loads, integrates communication and integrates sensor and actuators for performing the advanced monitoring and control features.

Electric vehicle energy usage information and electric grid status need to be collected using wired/wireless sensors to determine the efficient and economic charging operation of the EVs. During a trip the EVs driving patterns and the effective management of charging/backfill operations can be used to update electricity rates and flatten electric load curve due to different grid stability/ reliability and geographical location.

The EVs architecture includes the use of onboard devices to allow the driver to receive advice or seek instructions to efficiently manage the EVs battery charging/backfill process. Sensors will be integrated in the battery pack to detect when the battery capacity is below a threshold level. This will be connected with the GPS position of the vehicle and a list of near-by charging stations' locations. Based on the distance, the current prices and projected energy cost, the use of intelligent cloud computing will direct the driver with the optimum course of action.

The EVs batteries can be used to serve as an energy storage which can backfill into the local electric grid when not in driving status. This can help prevent power outage during peak demand. In this case, sensors are used to detect or predict instability in the grid and inform the driver to bring the vehicle to the appropriate charging station to serve as backfill battery. The EVs onboard chargers and fast charging stations usage patterns will be monitored using sensors and wireless communication modules that communicate with the vehicle and the infrastructure.

2.3 Communication

Communication plays a crucial role in enabling the full functionality of an intelligent distribution grid in that the overall control requires a mesh of sensor and actuators monitoring the state of the low voltage grid and able to take full advantage of an "internet-like" approach.

One of the primary objectives of the IoE project is to develop inexpensive, reliable communication hardware which is able to transmit the information across long distances in a secure way. Here, an industry-hardened variant of the power line communication (PLC) appears to be in many cases the technology of choice because of its proven ability to transport sensors information and actuators commands over existing low and medium voltage lines, without needing any complex and expensive installation of new wire.

PLC is already massively adopted by major energy Utilities, such as ENEL, A2A, ACEA, ENDESA, IBERDROLA, ERDF and many others to remotely access customer power meters and built an Advanced Metering Infrastructure (AMI). In this "Access" domain, some narrowband PLC technologies with baud rates ranging from few kbps to few hundred kbps have been widely accepted and deployed, as by the facto standards, supported and promoted by leading industrial Associations such as Meters&More and PRIME Alliances, others have been just proposed for new standards evolutions such as the IEEE P1901.2.

In the Home Area Networking (HAN) domain, RF technologies like ZigBee or M-BUS, are often used for energy management applications, but PLC is also becoming a good complementary technology to assure 100% communication coverage over the entire building. HomePlug standards are the most used PLC technologies in HAN domain with over 65 Millions devices installed to date. In HAN domain, by exploiting the broadband frequencies, HomePlug technologies may cover either Smart Energy specific applications (HomePlug GP) with baud rates of few Mbps or consumer applications like Home-Video and Internet distribution inside the house (HomePlug AV/AV2) with baud rates of some hundred Mbps. PLC is also suitable for V2G communication according to ISO/IEC 15118.

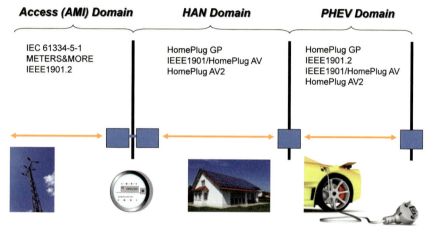

Fig. 2.   Smart Meter Domains - PLC Technologies Suitable for Smart Grids

2.4   Renewable Energy

The grid remains stable only when consumption and generation are balanced, and this becomes difficult when wind turbines or solar panels are connected. The wind fluctuates between gentle breezes and powerful storms, the sun is shining or the clouds are covering the sky and affect directly the energy generation of wind turbines or solar panels.

Today, in many European countries the grid operators are required by law to give priority to clean forms of energy when feeding electricity into the grid. The sun and the wind are very unpredictable and the energy fluctuations result in that the grids are reaching maximum load more and more often. In these conditions the modern power grid must be capable of integrating fluctuating loads into the existing system. The power companies are required to incorporate a multitude of small and very small energy sources. Homeowners are turning into producers of electricity as they install solar panels on their roofs and cogeneration plants in their basements. The grid becomes larger, more flexible and more intelligent. In this respect the use of smart solar array management would be crucial for the full utilisation of the generation capabilities.

2.5 Distributed Storage and Storage Farms

Batteries are generally not well suited to buffer the highly volatile renewable energies, even though they are able to store considerable amounts of electric energy. This is due to the fact that the batteries' electro-chemical working principle sets a limit to charging respectively discharging rates. Trying to exceed this limit, which is given by the fundamental principles of the device, will lead to functional deterioration, fast ageing or ultimately damage. In the light of today's battery prices, this is unacceptable and would constitute a road blocker if not avoided by alternative technical measures.

Fast storage devices contribute to a great extent to the stabilization of the distribution grid under the conditions of highly volatile energy production. The consortium therefore investigates storage alternatives with high dynamics such as arrays of supercapacitors, electro-mechanical flywheel principles and several more. All these approaches will be assessed with respect to their storage capabilities, dynamics and long-term reliability. The latter may be achieved by using proven components from the electric vehicle infrastructure. Other vital aspects are safe and secure operation and low cost at high volumes. Special emphasis will also be on modular concepts which allow stacking of devices for flexible scaling of the maximum stored amount of energy.

3 Energy Utilities

The energy utilities base their energy gross market purchases on the prediction of the consumption according to statistical models and must offer the amount of electricity that people and industry need at any given time. Regardless of fluctuations of gross prices, the utilities charge a largely uniform price for that electricity (in some countries some differences in between the peak and the off peak hours have been introduced recently).

The introduction of distributed renewables is now requiring the fluctuating consumption to be adjusted according to the fluctuating supply and the variable energy prices. This requires new business models and variable pricing models in principle similar to those that are now common in the telephony market. In this new environment the forthcoming electric vehicle owners would appreciate the opportunity to select electricity providers on a case-by-case basis (just like having several rechargeable SIMs, or mobile phones available and using pre-paid tariffs), or they could simply buy monthly flat-rate packets of kilowatt hours at the preferred market prices, which would essentially be similar to a post-paid telephony contract.

The technology aspect and the smart devices which are required and are current under development for the implementation of these market models are essentially the demand-production models, the smart metering devices, the smart home automation appliances and, for the electric vehicle and the distributed home RE plants and storage the smart bi-directional chargers-inverters.

3.1 Grid Stability, Matching Demand and Production

A massive restructuration of the distribution grid is currently taking place in Europe both from the technological and the regulation point of view: actions are being implemented both at a national and at a European level as shown in Fig. 3.

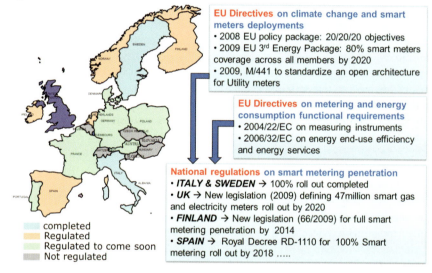

EU Directives on climate change and smart meters deployments
• 2008 EU policy package: 20/20/20 objectives
• 2009 EU 3rd Energy Package: 80% smart meters coverage across all members by 2020
• 2009, M/441 to standardize an open architecture for Utility meters

EU Directives on metering and energy consumption functional requirements
• 2004/22/EC on measuring instruments
• 2006/32/EC on energy end-use efficiency and energy services

National regulations on smart metering penetration
• *ITALY & SWEDEN* → 100% roll out completed
• *UK* → New legislation (2009) defining 47million smart gas and electricity meters roll out by 2020
• *FINLAND* → New legislation (66/2009) for full smart metering penetration by 2014
• *SPAIN* → Royal Decree RD-1110 for 100% Smart metering roll out by 2018 .....

completed
Regulated
Regulated to come soon
Not regulated

Fig. 3.   European and National Regulation on Smart Metering

Up to now the grid stability was dominated by the concept of the frequency synchronisation across the network in between the alternator turbines and the final user's loads (essentially resistive loads and AC 50Hz electric motors) and therefore the grid stability was assured by maintaining a constant frequency and a constant $cos\phi$.

The new operating conditions which see a predominance of electronic loads is making this concept completely useless: the mains frequency as a control variable for the total power consumption is loosing its meaning. Instead, the operators need to determine in real time the load state of every single generator and the load within the various grid partitions in order to establish operating conditions under which grid stability can be ensured. In addition they must rely on complex databases and prediction models for the real-time prediction of the characteristics of the load in order to anticipate the variability that can occur without loosing the control of the network.

This inherently relies on communication systems and intelligent strategies to allow the highest degree of utilization of the power plant productivity. When communication fails, fall-back strategies at local and global level must be enabled.

The IoE project is devoting special emphasis on the investigation of technical measures, algorithms, embedded systems and new concepts for electronic hardware components which assure the same degree of grid stability even in the case of communication failure.

3.2 Smart Meters

One of the technologies supporting the new smart grid is the new smart meters connecting the grids with the consumers like the electric vehicles. The new smart meters record all data in real time and can inform the driver how much energy is used, and how and when it is used. This information helps the electric vehicle energy management system to make better decisions, such as when to use energy to find lower rates and when to switch to more energy efficient modules. In much the same way that the Internet changed the global economy and transformed the business landscape, smart meters and the smart grid combined into the Internet of Energy can change the way the users consume energy. This will unleash a wave of economic growth.

The smart energy meter will communicate, directly or through a "communication HUB", with the surrounding infrastructure devices (i.e. the Internet of Things, e.g. the vehicle-to-vehicle or the vehicle-to-infrastructure network) or in a vehicle-at-home scenario (i.e. cradled and therefore connected to the high-speed network), to send/get real time pricing, mission history and predictions, status of health data to/from the end user energy consumers.

Smart metering devices in the Internet of Energy scenario will be embedded in EVs, home appliances, charging stations etc. They will measure many electrical parameters, such as max and min power demand, current, voltage, power factor, and they will communicate through embedded power line modems or other communication technologies with different aggregators (from the home server to the utilities' data farms) providing information about e.g. power outages.

New available super-integrated flexible and scalable power line communication platforms can be adopted to address different smart grid application needs and protocol standards in this case.

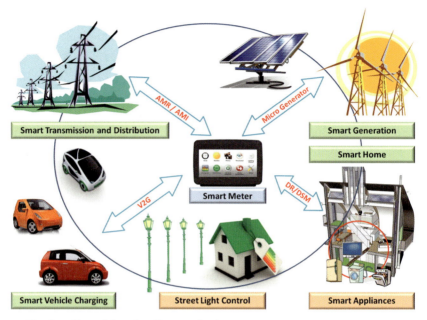

Fig. 4.   Smart Metering Communication

The new "STarGRID" PLC platform from STMicroelectronics is for example able to integrate in a single chip an optimized DSP engine for different modulation schemes (FSK, S-FSK, PSK, OFDM) a digital 8-bit industrial microcontroller core for system supervising and multiple narrowband protocol stacks management (IEC61334-5-1, Meters&More, PRIME...), the full receiving and transmitting analog front end and an integrated power amplifier with on board programmable filtering, capable of more than 1A (RMS) of current to drive very low impedances loads, so offering the needed flexibility with the highest integration level and lowest power consumptionA 128-bit AES data encryption engine is also optionally available to guarantee customer data authentication and privacy, being always smart grid security a key element to be taken into account. The "STarGRID" PLC platform is presented in Fig. 5.

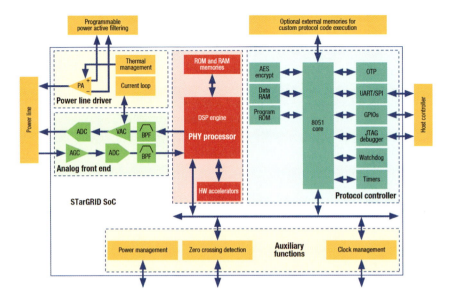

Fig. 5.   "STarGRID" PLC platform

3.3 Smart Appliances

The grid will increasingly rely on smaller, locally distributed electricity generators and storage systems that are based on plug and play principles. Power network devices and loads at the edge, such as electrical vehicles, buildings, electric devices, and home appliances, can be charged or connected on any source of energy being solar, wind, or hydroelectric.

Reference designs and ESs architectures for high efficiency innovative smart network systems are developed with regard to requirements of compatibility, networking, security, robustness, diagnosis, maintenance, integrated resource management, and self-organization. The accelerated fast bidirectional onboard chargers, the different gateways including smart metering and on-vehicle power management features accessible by internet communication gateway will communicate with the smart appliances and will use common gateways.

3.4 Bi-Directional Chargers/Inverters

To ensure compact design and low electric power loss the Internet of Energy concept connected to the electric mobility necessitates the availability of bidirectional inverters with very high energy densities, excellent efficiency and state-of-the-art thermal management.

Charging an electric vehicle (EV) equipped with 400 volt technology and a battery capacity of 20 kWh in a time interval of 200 minutes (roughly 3 hours) entails a power rating of about 7 kW – roughly equivalent to an electric cooker at full operation. Beyond that the capability of bidirectional operation is prerequisite to transfer electric energy from the battery of the EV into the distribution grid. The higher the rated power of the charging device the faster is the charging process.

Drive inverters might be used additionally as charging devices. Even more so, it is required that the inverter features a filter function (e.g. for reactive power compensation) in order to improve the power quality. A further beneficial feature is the grid stabilization capability. Thousands of EVs can support the grid in a decentralized grid topology.

Last, it has to be equipped with an interface to a residential energy manager which operates as an energy broker in order to optimize the costs for charging the EV according to the users' demand. Additionally the power flow to and from the battery will be optimized by strategies and cost information provided by the residential energy manager.

The matter of investigation within the framework of the IoE project will be to find a set of technologies which allow the realization of such requirements at reasonable costs for such a presumable mass product which is likely to be found in the majority of households. In this context, starting from proven technologies, also new semiconductor basis technologies for the inverter components and promising topologies will be investigated and tested.

3.5 Fast Chargers

To charge a fully electric vehicle within a timeframe of 20 minutes requires an inverter with a power rating of about 60 - 100 kW and a current rating of about 230 Amps. Even more than in the case of the bidirectional chargers, breakthrough concepts have to be developed for unprecedented power density, low-loss architecture and outstanding thermal management in order to minimize the efforts for cooling, losses and cost. Especially in this case, new semiconductor technologies on the basis of Silicon Carbide (SiC) components will be investigated and tested in order to fulfil the highly challenging requirements.

4 Energy Consumers

4.1 Electric Vehicles

EVs are expected to play an important role in the smart grid (SG) and IoE concept. Electric-drive vehicles (EDVs) include battery electric vehicles, fuel cell electric vehicles and hybrid electric vehicles [3]. In our paper we exclusively deal with the plug-in enabled EDVs. EDVs incorporate batteries which are capable of storing a significant amount of energy. Summing up all the possible electrical energy stored in the expected amount of EVs in near future, one finds that it is possible with a constructive bidirectional interaction between the EVs and the upcoming smart grid. Many new EVs are now entering the automotive market, and a large increase of EVs on the roads is anticipated in the coming years. As sales numbers rise, so does the competition between the EV producers. This in turn results in new innovative developments, e.g. in control systems, technologies, devices and products.

The idea of a bidirectional connection between the EVs and the grid is not new. Kempton and Lentendre [4] suggested a "two-way, computer-controlled connection to the electric grid". In this way they claim it would be possible that the grid both provides power to and receives power from the EVs.

Power transferred from the vehicle into the grid is commonly named "Vehicle-to-Grid" power (V2G) [3]. In addition to be a power resource for the grid during demanding periods, the EDV fleet can act as storage for large-scale wind power [5] and other intermittent energy sources, e.g. solar and wave. Higher grid flexibility and control has to be implanted as these large-scale variable energy sources enter the market and connect to the existing grid.

The availability of sufficient storage capacity for electric energy is a pivotal prerequisite for the integration of highly volatile electric power sources like wind turbines and photovoltaic devices. While at present several European hydro power plants are going to be reconstructed for pumped storage [6], this might not be the optimal solution in a future scenario where electric power distribution is organized as an array of decentralized micro-grids on the scale of residential areas or villages. A much more favourable and intriguing solution would be the incorporation of full electric vehicles, not only for power consumption, but also as power providers. During energy generation peak periods, they can store excessive energy in their batteries when connected to a charging station and feed it back to the grid in times of high power consumption demand.

The potential of this approach becomes evident looking at the annual number of passenger car registrations in the European Union. During the last five years, the average amounted to about 13 million cars [7]. With a storage capacity in the 20 kWh range for a typical EV - which is expected to increase with growing energy density of batteries - this would amount to a total annual growth of about 260 GWh. This is equivalent to the typical annual consumption of 60,000 4-person-households [8] if all these cars were full electric vehicles!

The significance of such a storage capacity becomes even more apparent if we compare it with that of an up-to-date pump storage power plant. The well-known Kaprun power plant has lately been expanded by the pump storage facility "Limberg II" [9] which is located at an altitude of 369 m above the existing dammed lake. Its useful capacity amounts to 84.9 million m^3. Neglecting the degree of efficiency of the turbine and its electric installation, the potential energy stored in this new artificial lake amounts to 85.4 GWh. Again assuming the electric energy consumption of the 4-person-household, about 19.000 households could be supplied with electric energy for one year. 13 million full electric vehicles per year would mean an additional storage capacity of more than three Limberg II pump storage power plants!

Important topics to address in order to enable the realization of the V2G are related to the grid, the EDVs and the interface in between the two. Advances are being or expected to be made in the following areas:

- ▶ Grid interconnection to the Internet for application such as electric mobility
- ▶ Hardware, software and middleware development for seamless, secure connectivity and interoperability achieved by connecting the Internet with the energy grids
- ▶ Communication circuits
- ▶ Embedded systems modules for control and monitoring
- ▶ Smart Grid Remote Controls
- ▶ Microsystems technologies
- ▶ SiC
- ▶ Power electronics components
- ▶ Electrical interfaces
- ▶ Plugs
- ▶ Communication
- ▶ Underlying architecture
- ▶ Distributed Embedded Systems (ESs)
- ▶ Component integration combining power electronics integrated circuits, sensors, processing units, storage technologies, algorithms, and software
- ▶ Application of the IoE is the infrastructure for the electric mobility

▶ Charging requirements
▶ Standard Charging
▶ Fast Charging
▶ Bidirectional Charger for V2G
▶ Controls
▶ Converters

5 Summary

The Internet of Energy concept is defined as a dynamic network infrastructure based on standard and interoperable communication protocols that interconnect the energy network with the Internet allowing units of energy (locally generated, stored, and forwarded) to be dispatched when and where they are needed. The related information/data will follow the energy flows thus implementing the necessary information exchange together with the energy transfer. Building the Internet of Energy will be the answer to a number of the energy challenges related to the implementation of the infrastructure for the electric mobility and the full deployment of the renewable energy production, while the advancements in nanoelectronics, microsystems, embedded systems, communications, control, algorithms, software and Internet technology addressed in the IoE project are the enablers that will make the implementation of the concept possible.

References

[1] Asplund G., Sustainable energy systems with HVDC transmission, IEEE Power Engineering Society Meeting 2004 proceedings, pp. 2299-2303, June 2004.

[2] http://en.wikipedia.org/wiki/List_of_HVDC_projects

[3] Letendre, S. E., Kempton, W., The V2G Concept: A New Model For Power?, Public Utilities Fortnightly, pp. 16-26, February 2002.

[4] Letendre, S. E., Kempton, W., Electric Vehicles as a New Power Source for Electric Utilities, Transpn Res.-D, Vol. 2, No.3, pp. 157-175, 1997.

[5] Dhanju, A., Kempton, W., Electric Vehicles with V2G Storage for Large-Scale Wind Power, Windtech International, 2006.

[6] http://www.verbund.com/cc/de/news-presse/aktuelle-projekte/oesterreich/limberg-2

[7] http://epp.eurostat.ec.europa.eu/portal/page/portal/transport/data/database

[8] http://www.verivox.de/nachrichten/starke-unterschiede-beim-stromverbrauch-in-europa-12274.aspx

[9]   Verbund AG (ed.), „Saubere Energie für Generationen – Pumpspeicherwerk Limberg II" Verbund AG, October 2010.

Ovidiu Vermesan, Lars-Cyril Blystad
SINTEF
P.O.Box 124 Blindern
0314 Oslo
Norway
Ovidiu.Vermesan@sintef.no
Lars-Cyril.Blystad@sintef.no

Reiner John
Infineon Technologies
Am Campeon 1-12
85579 Neubiberg
Germany
Reiner.John@infineon.com

Roberto Zafalon, Alessandro Moscatelli
STMicroelectronics
Via Olivetti, 2
20041 Agrate Brianza, Milano
Italy
roberto.zafalon@st.com
alessandro.moscatelli@st.com

Kai Kriegel, Randolf Mock
Siemens AG
Otto-Hahn-Ring 6
81739 Munich
Germany
kai.kriegel@siemens.com
randolf.mock@siemens.com

Marco Ottella, Piero Perlo
Centro Ricerche FIAT S.C.p.A.
Strada Torino 50
10043 Orbassano
Italy
marco.ottella@fptpowertrain.crf.it
pietro.perlo@crf.it

Keywords: internet of energy, smart grid, electric vehicle, renewable energy

Efficient Allocation of Recharging Stations for Electric Vehicles in Urban Environments

J. Gallego, E. Larrodé, University of Zaragoza

Abstract

With the aim of decreasing today's dependency on oil, a growing interest in vehicles powered by alternative fuels, such as electric vehicles, has been generated. However, the successful incorporation of electric vehicles in the actual transportation system depends on overcoming of two main barriers: the low range of the electric vehicles in comparison with traditional internal combustion vehicles, and the low number of electric recharging stations. By increasing the number of recharging points it is possible to compensate the range limitations of electric vehicles, nevertheless, creating this infrastructure causes high costs. Therefore, the objetive of this study has been to develop an optimization methodology that allows planning of the recharging infrastructure for electric vehicles in an urban environment that minimizes the cost.

1 Introduction

The internal combustion engine is the propulsion technology that traditionally has been used in vehicles. However, motivated on the one hand by the end of the cheap oil as a result of the excesses committed in the past and the lack of foresight, and the higher demand caused by an economic growth of countries under development like China, India or Brazil [1], and on the anoter hand motivated by socio-political aspects, dependence on imported oil from third countries, and environmental problems, such as greenhouse gas pollution and local pollution in cities [2], have generated growing interests in vehicles powered by alternative fuels. Alternative fuel vehicles include hydrogen, biodiesel, natural gas, ethanol and electric vehicles.

The successful incorporation of these alternative vehicles into the road raises some difficulties: limited numbers of refuelling stations, high refuelling costs, onboard fuel-storage issues, safety and liability concerns, improvements in the competition and high initial costs for consumers [3].

In particular, before obtaining a successful incorporation of electric vehicles in the actual transportation system it is necessary to recover two crucial aspects: the electric vehicles's range and the development of a suitable and dispersed recharging infraestructure. By increasing the number of recharging points it is possible to compensate the range limitations of electric vehicles [4], nevertheless, this infrastructure creates high costs.

Many studies have been developed with the objetive of achieving exact or approximate methods (heuristics and metaheuristics algorithms) that minimize the cost of developing this infrastructure by optimizing the location of refueling stations. This optimization has been based on the range of the vehicles and major vehicle flows to simultaneusly serve inter city and intra city travel. Two of the more popular approaches for optimal location of alternative-fuel stations are the p-median and the Flow Refueling models.

The p-median model is a location-allocation model that locates a given number of facilities and allocates demand nodes to them to minimize the total distance travelled by consumers to facilities. This model locates Satations conveniently to where people live or work. Lin et al [5] treat the station sitting problem as a fuel-travel-back problem, a typical transportation problem, and solve it using a mix-integer-programing model based on the distribution of vehicle miles traveled; it is structurally similar to the p-median problem.

The Flow-Refueling Location Model (FRLM), introduced by Kuby and Lim [4], is a path based demand model that locates a particular number of stations to maximize the number of trips with refuelling at the shortest path. This model counts a flow as refueled only if a combination of existing stations on a path that can successfully be used to refuel the round trip between the origin and destination without running out of fuel or electricity, given the assumed driving range of vehicles. Kuby and Lim developed a mixed-integer linear program (MILP) formulation of the FRLM that has been used in some studies; i.e. Wang [6] used it to optimize the location and number of battery exchange stations, or Wang and Lin [7] have applied it to focus the model on facility sitting to cover the passing flows along the shortest paths of interest, instead of sitting central facilities to serve demand at determine points. However, the MILP only is effective in obtaining optimal solutions for small networks; with the aim of solving FRLM for realistics networks within a reasonable timeframe it has been necessary to develop heuristic algotithms, including greedy-adding, greedy-adding with substitution, and genetic algorithms [8].

Upchurch and Kuby [9] compare the node based p-median model and the flow-based FRLM and conclude that the FRLM facility locations perform better on the p-median objective than the p-median facilities do on the FRLM objective, in other words, the FRLM does a better job fulfilling stations near people's homes and stations convenient to people's trips.

In this research the objetive is to develop a tool to obtain the plannig of the recharging infrastructure for electric vehicles in an urban environment, taking into account aspects not considered previously, such as the size of the recharging station, the technology or the available plots within the city; in this case, the range achieved by vehicles is not a factor as decisive as in previous studies because of the distances between stations are shorter than in inter-city trips. The tool focuses on finding solutions for those particular cases in that is necessary to give efficient recharge service (availability, fast recharge) to private fleets and vehicles. The case of slow recharge on housing for private vehicles, despite being the type of recharge is expected to extend more quickly, is not going to be considered because in the early stages of implementation it is sufficient to consider the incorporation of accountants in the existing electrical network of the garages.

2   Methodology

Development of this tool will help to determine in the most efficient way, the number and location of the electric recharging points, as well as the station sizing, according to the different demands at different temporary scenaries. This choice of the optimal location, which regards to logistics and technical criteria, has a great importance due to the high cost that supose to create these facilities create.

Nowadays, there are not any criteria of optimization to locate these kinds of facilities. In case of localized fleets, their location follows proximity criteria, and in many cases these facilities are for private use. That is why if the use of electricity in transportation was general and those facilities were opened to the public, the number of stations would be insufficient and their exploitation would present serious logistical operation problems. This issue is expected to be solved with the development of this tool.

Planning should be guided by a methodology based on studies of mobility in a determinated geographical environment. This geographical environment must allow determining the necessary number, size, in function of the number of vehicles which will be served, the kind of these vehicles, the use and rules and standardization, and the allocation of the recharge points to install within the city, in function of the vehicles characteristics paths, the location of the main residential and workplace areas, logistic platforms location, grid limitations, traffic densities and legislation. Also, the location of different recharging points will be made according to cost minimization criteria. The new concepts of electricity supply to fleets will be studied and necessary technologies for non-polluting electric vehicles will be collected. Due to the physical space required by new devices and the new infrastructure, is going to particular attention will be paid to energy storage systems such as battery packs or on site energy renewable sources. Previous considerations for the development of the tool are that electrical demand and elements included in the future recharging stations are known, and the choice of final configuration is done according to meet demand criteria regarding electric power supply, cost and available space. On the other hand, aspects such as the electrical network design and facilities design and architecture are not taken into account. Fig. 1. shows the methodology followed for the tool developed. This methodology includes decisional algorithms. The following explains the operation.

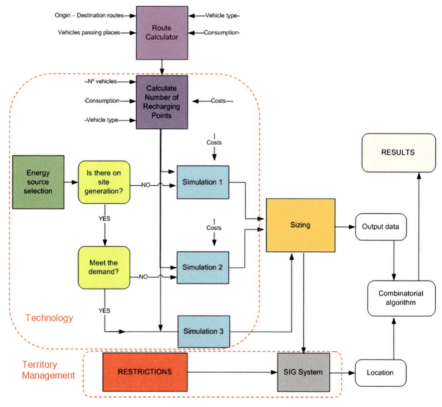

Fig. 1.   Methodology Diagram

2.1. Routes

The tool provides the opportunity to conduct a study of the routes used by vehicles in order to know the consumption profile and thus know the electrical demand of those vehicles or fleet which are expected to be users of the electric recharge station. The knowledge about the amount of electricity demanded in a determined period of time helps to get a close-fitting facility sizing.

Fig. 2.   Route Calculator

2.2 Recharging Points

First of all is necessary to determine the number of recharging points required to meet the total energy demand. In this sense, indispensable data is to know the expected electrical demand. This can be achieved by, either an estimation from the fleet and its features that is expected to be the user of the facility, or more accurately by the routes calculator, from the consumption of certain vehicles that follow fixed routes.

2.3 Energy Source Selection

Next step is to select the electricity source. Once the number of recharging points has been determined and if there is an on site renewable energy source, it must be considered if this energy source meets the total energy demand or not. This is going to determine the way forward. There are three options:

▶ If no renewable generation exists on site: all energy is obtained from the grid (Simulation 1).
▶ If on site renewable generation exists: it is necessary to know whether the facilities have been sized to meet the electrical demand or not. If they are unable to meet demand, it is necessary to determine the energy to be obtained from the grid, regardless origin (Simulation 2).
▶ If on site renewable generation exists: it is necessary to know whether the facilities have been sized to meet the electrical demand or not. If they meet demand, is not necessary to obtain electricity from grid (Simulation 3).

2.4 Sizing

Depending on the desired design option, on-site renewable power generation, grid supply or both; on the one hand are indicating the mandatory elements associated with each option, and on the another hand give the chance to incorporate other optional devices. Knowing the mandatory and optional elements, as well as the minimum safety distances according to regulations, the number of recharging points and the area for maneuver and recharging in function of the types of vehicles that are expected to be the facility users, the minimum area required to house the facility can be estimated. Output data process will summarize the elements that include the global infrastructure.

2.5 Restrictions

The restrictions allow delimiting the optimum locations search among all possibilities selected by the combinatorial algorithm:

▶ Maximum number of recharging stations. The tool allows selecting the maximum number of stations to be build for meeting the demand.
▶ Minimum safety distances. It refers to the minimum distance with the surrounding environment.
▶ Stationary routes. The tool allows entering a determined route on a map. Then, when the optimum locations are determined, priority is given to those which are closest to these routes (especially if there are bus routes).

▶ Average Traffic Intensity (ATI). The tool allows entering to the main roads and intersections the average rate of traffic intensity. Then, when the optimum locations are determined, the busy areas are considered.

▶ Confluence areas. The tool takes into account those areas of greatest confluence of vehicles, for example, shopping centers, industrial zones, residential areas, workplaces, etc. These areas are where vehicles remain parked for longer, giving priority to them when optimum locations are selected.

▶ Restricted areas. Refers to areas in which based on local legislation or safety rules, is not possible to install the stations. The tool allows setting them on a map.

▶ Grid limitations. The tool allows to identify areas where is not possible to connect to the grid because of it is already at the limit.

2.6 SIG Location

In the previous step the maximum number of recharging stations has been determined. With the aim locating these stations on a map first of all it is necessary to check the availability of this maximum number of places that satisfy:

▶ Constraints set out in the restriction process.

▶ Availability and suitability of place and ground (especially important when on site renewable generation does exist).

In the cases that have been found the maximum number of places that meet all conditions, it is necessary to determine the exact location and calculate the total cost. If not, is necessary to adjust the number of possible stations. The total electrical demand is divided for one recharging station less than in the previous case, and the process is repeated. This will be held until a solution is found. Finally, the tool must graphically return the optimal solutions through an SIG application (Geographical Information System). In this case has been used ArcG is used, an application that creates, shows and uses maps.

2.7 Combinatorial Algorithm

The output data obtained from all the options generated in the sizing process and all the solutions obtained from the Territory Management module, are introduced into the Combinatorial Calculation Algorithm. The number of combinations is too large; therefore, in order to optimize the searching process, the resulting combinations will be explored through a guided discussion based on heuristic algorithms. The tool uses a model for optimal location that include variations with the aim of taking into account all constrains considered in the study.

Fig. 3.   Location analysis created by a SIG system

The tool allows two ways of working:
- ▶ Option 1: Several models are manually selected for each device that exists in the facilities, then, they are combined to get the optimal solution.
- ▶ Option 2: Selection criteria are used (cost, size, accessibility). Thus it is possible to assign priority values to each criterion, such that the tool will order the obtained solutions according to the primary selection criterion.

3   Results

With the aim of analyzing the different solutions obtained and to determine the optimum one, there are three different criteria:
- ▶ Solution total cost. It considers the cost of equipment, energy, grounds, maintenance, operation, preparation, implementation and building.
- ▶ Total area required by the total number of recharging stations to meet demand.

▶ Efficiency.

It is possible to determine: the number of facilities to meet demand, the cost, the required area, the power source and models for the necessary equipment.

3.1 Power Supply Facilities

The tool provides information about the recharging stations. These are the principal part in the future electrical vehicle supply infrastructure. The results obtained are the following:
▶ Demand characteristics of the recharging stations: frequency of consumption, time curves, temporal statistics.
▶ Number and characteristics of the required recharging stations.
▶ Location.

In this context, the tool must be able to provide the following outputs: the plot dimensions and its location on a map.

3.2 Costs

In optimal location problems, costs are one of the determining factors in making decisions. The maintenance, operation, equipment, on site power generation, grid energy, plots, implementation, insurances, guarantees, permits or licenses, are those costs that must be addressed in the decisions making process. Not all of them will be present in all the recharging station configurations, but they will depend on the particular characteristics of power generation or energy supply systems.

4 Conclusions

The methodology presented here takes the following variables into account: an overview of the characteristics of the electric vehicles that are going to be used, and an overview of the environment under study: density of vehicles and principal routes, analysis of origin-destination, safety and standardization, and grid limitations. With this tool is possible to obtain the optimal location and cost, the determining factors, of a recharging network, as part of the future electrical infrastructure. This methodology allows us to realize simulations taking by reference different scenarios of electric vehicle penetration. Thus, the results are an initial database for the further impact analysis about the electrical transport and distribution system.

References

[1] Krishna, V. Rapid economic growth and industrialization in India, China & Brazil: at what cost?, William Davidson Institute Working Paper Number 897, 2007.

[2] Silva, C., Ross, M. y Farias, T. Evaluation of energy consumption, emissions and cost of plug-in hybrid vehicles, Energy Conversion and Management, vol 50, 1635 - 1643, 2009.

[3] Romm, J. The car and fuel of the future. Energy Policy, vol. 34 (17), 2609 - 2614, 2006.

[4] Kuby, M., Lim, S. The flow-refueling location problem for alternative-fuel vehicles. Socio-Economic Planning Sciences, vol. 39, 125-145, 2005.

[5] Lin, Z., Ogden, J., Fan, Y., Chen, C. W. The fuel-travel-back approach to hydrogen station siting, International Journal of Hydrogen Energy, vol. 33, 3096 - 3101, 2008.

[6] Wnag, Y.W. Locating battery exchange stations to serve tourism transport: A note, Transportation Research Part D, vol. 13, 193 - 197, 2008.

[7] Wang, Y. W., Lin, C. C. Locating road-vehicle refuelling stations, Transportation Research Part E, vol. 45, 821 - 829, 2009.

[8] Lim, S., Kuby, M. Heuristic algorithms for sitting alternative-fual stations using the Flow-Refueling Location Model, European Journal of Operational Research, vol. 204, 51 - 61, 2010.

[9] Upchurch, C., Kuby, M. Comparing the p-median and flow-refueling models for locating alternative-fuel stations, Journal of Transport Geography, vol. 18, 750 - 758, 2010.

Jesús Gallego, Emilio Larrodé
Universidad de Zaragoza
Campus Río Ebro, Edificio Betancourt
C/María de Luna s/n
50018, Zaragoza
Spain
jgallego@unizar.es
elarrode@unizar.es

Keywords: electric vehicle, recharging stations, urban environment, methodology, location, optimization

EPV Project: ICT Tools & EV Integration Inside Smart Grids

P. Peral, S. Santonja, Energy Technological Institute
A. González, Iberdrola Distribución Eléctrica, S.A.

Abstract

This paper presents the advances of the EPV (Electrically Powered Vehicle) Project, which searches the analysis and design of specifications of a new energy efficient urban transport system. In these days where Information and Communication Technologies (ICT) play a key role in our society and the environmental concern is increasing, opportunities of linking both issues are arising, which will help to make a profitable use of clean, renewable energy resources. The technological goal of the EPV Project is to obtain a complete development of the hardware and software to optimize infrastructures and to develop new communication architectures. Introducing the concept of electric vehicle (EV) as a smart load in the intelligent network, the project is aimed to identify the possible integration of renewable energies in the EV charge, modeling and analysis of the distribution network for EV penetration.

1   Introduction

The EPV Project was launched as a scenario with an increased interest for the EV that requires an optimal infrastructure design in different levels and a bidirectional communication between the vehicle and the management agents. In the main goal is to obtain an intelligent and efficient distribution (Smart Grid) from the sustainable resources, which are around Europe, to the final users, optimizing energy storage and the use of ICT in order to provide the traffic management with new tools. The transport electrification is growing and changing the urban traffic patterns. The power grid should be able to support so many connections at the same time, managing also so different charge profiles as vehicle drivers. Therefore, the Information and Communication Technologies must be implemented in the development of this integration. Nowadays, navigating systems give drivers streets status, optimized routes, trip time, etc. In a word, they give important and complete knowledge about their trip.

In order to decrease CO_2 emissions, the development of a transport system based on EV provides the EPV Project with a real tool to improve customers trips, adapting the technology to their needs to reach a similar performance of that of the transport systems used today. Also, the EPV Project works on the analysis of customer driving and habit patterns to identify the possible EV penetration, standardization of charging station pods, control software for a business model that integrates generation and demand management, and also, how to add new devices to the daily life carefully, in order to provide a favorable answer from the final potential users.

Considering all these issues, the EPV (Electrically Powered Vehicle) project was launched, financed by Valencian Government and the European Regional Development Fund (ERDF). The objective of the project is the development of a new efficient transport system based on electric and hybrid vehicles, integrated on the grid with high renewable energies integration.

With the advice of the Valencian Energy Agency (whose Spanish acronym is AVEN), the Energy Technological Institute coordinates the project with a consortium of nine leading enterprises in different sectors (Iberdrola Distribución Eléctrica S.A., Movilidad Urbana Sostenible, Power Electronics, Nutai, GND, Tecnibat, CPD, IDOM and Montesol), to identify consumption patterns, new regulatory models and charging infrastructure for the electrical batteries. Also, Iberdrola and other partners will analyze the use of wind power to charge the EV fleet, adapting the energy demand associated with this type of vehicle to the periods of maximum generation.

2 Modeling and Analysis of the Power Grid

Looking at the intelligent network proposed, the EPV Project aims to reach the European Union is objective for the distribution model as pointed out in the Transport White Book [1]. The studies performed will analyze the impact of the EV connection on the grid, how this affects the main issues of the electrical system, and the interaction with a wind power generation source. However, defining user profiles of electric vehicles and their needs for mobility and charge is the first task.

2.1 Customer, the Most Favored

"Customer is always right". This expression puts the costumer in front of the first column of studies on the EV integration in smart grids. The active involvement of customer improves the project results, depicting the best way to implement new technologies in vehicles and charge station sequences. In this way, with this new sustainable transport system, the trips inside the city and to the working centers would be more energy efficient, reducing traveling expenses, CO_2 emissions and noise in urban areas.

Our study case is the metropolitan area of Alicante, at Valencian Region, with a population of 452.500 inhabitants (2008). For the achievement of the work package objectives, we have worked with the following sources: Mobility inquiry of Alicante in year 2001, Movilia 2007 survey. Survey conducted by the Ministry of development, Socio-economic data by Statistics National Institute (2008) and inquiry at Alicante Airport. Some driver patterns have been identified, paying attention to those ones which cover about 70% of population. A big group denominated "Commuter" has been detected with some specific characteristics, being short trips of around 30 minutes and 150 km per day, specific trips (usually home to work and work to home), urban movements and fewer trips during the weekend. From these analyses, optimal charge station locations are being obtained.

2.2 Electric Vehicles Penetration with High Wind Power Integration

Different analyses are being performed on the network, applying the PSS/E software, with the objective to analyze the impact of the EV connection on the grid, how it effects to the main issues of the electrical system, and the interaction with a wind power generation source. Our scenario is the analysis of the Iberdrola Distribución S.A. medium voltage network in the metropolitan area of Alicante (Spain) (see Fig. 1).

As a result, the future network will combine the management of wind energy generation and the energy demand from the selected charging stations through the electrical medium voltage distribution network.

These analyses include power flow, fault and contingency analysis. Load flow analysis is the most important and essential approach to investigating problems in power system operating and planning. This is helping to design a model that will consider if the current network is ready to support the new electric vehicles load or not, and which networks conditions must be met to reach this condition. For the optimal location of charging points for electric vehicles, an analysis is performed to choose the secondary substations from the distribution network which would best fit to the necessary requirements. For this purpose, multiobjective evolutionary algorithms [2] are being used, which produce optimal configurations according to electrical, geographical and population habits criteria (see Fig. 2.). The application is being developed in Python language as an interface to the PSS@E, software, from where the network characteristics are read, and the power flow analyses are performed.

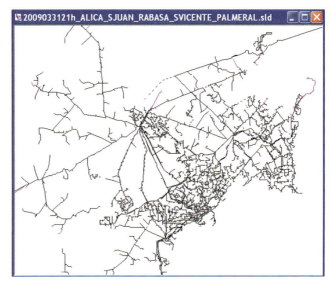

Fig. 1.   Iberdrola Distribución Power distribution network in the metropolitan area of Alicante

Some of the criteria to optimize being taken into account are: distances among charging points, voltages at nodes of the network, amount of trips being generated or reaching destination at each area, and overload on transformers (see Fig. 2.).

3   Charge Technology and Demand Management Tools

Technologically, the project tackles the developments at the utilities and charge station manufacturers side. In this way, it adresses tasks highlighted in the other columns of EV integration.

The EPV Project proposes a complete communication architecture, taking into account the penetration studies, profiles identified and the optimal location for charge station pods obtained by previous analyses.

3.1 Intelligent Charge Station (ICS)

For this project it is very important to achieve the complete integration of efforts coming from utilities, retailers and pod designers. It is considered the best way to get control over the charge, improving the electrical vehicle penetration and renewable energy resources use. ICT integrated in the developments and good designs of the control centre software are necessary to accomplish this goal.

First, the intelligence level needed at the charge station pod electronics has been considered. Taking into account the UNE-EN 61851-1 mandate [3], amount the device applied at the charge point must be able to storage a huge number of data (at least during a week), and also ensure a correct and safe connection of the vehicle, the bidirectional transfer of data with a Control Center, etc.. During the first stages of the project, the specification of the station pod has been defined to include all this requirements, searching the best design to obtain a modular architecture that makes future V2G communication easier.

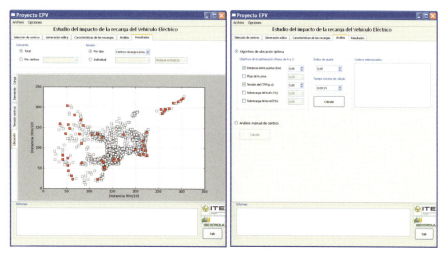

Fig. 2.   Distribution of Secondary Substation obtained in the algorithm Application developed at EPV Project and Criteria selection in the Algorithm Application developed at EPV Project.

In this way, the first selection was the connector between intelligent charge point and the vehicle. Although the specifications are in constant change and the current infrastructure is not prepared to admit this kind of communication, the Mennekes plug-in device [4] has been chosen as an appropriate connector in the EPV project. This connector provides the status information of the recharge through digital and analog signals (See Fig. 3).

In this way, the information could be transmitted to the customer, who increases his/her comfort by applying it in an optimized recharge (from the point of view of economy and also the time expected). In order to accomplish a functional prototype, which can be used in the current infrastructure of the power grid, it has been decided to develop an intelligent charge station with two different kinds of plugs. The first one is the indicated Mennekes plug and the second one with a standard Schuko plug. The last one, that runs in the range of 230V and 16A and do not provide any kind of communication, is the usual connector for domestic devices which allows to integrate the project prototype into the current urban areas that way at least the consumption impact and the customer answer can be analyzed. Furthermore, the "box" which will contain electronic and control devices, is being designed, looking at European connector standards and anatomic solutions. The first version of the "box" was developed with ecological materials, taking into account environmental concerns.

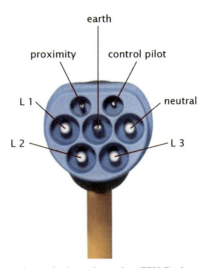

Fig. 3.   Mennekes plug solution adopted to EPV Project

3.2 Rapid Charge

An important element in the charge infrastructure is an energy storage device able to meet all the requirements defined. Furthermore for the development of rapid charge method, the performance of batteries can be expectedn to be reduced quickly. In the project, the specification and topologies expected to be used are discussed, concerning the rapid charging to provide the possibility of recharge the 80% of depleted batteries in a few minutes. Although this method implies a strong effort on the battery, some drivers would demand it during long trips or specific situations. Our first analysis will be with NiMh battery technology, checking current data in hermetic batteries [5] [6]. Anyway, Li-ion batteries are going to be tested, since they are a growing technology, together with the EV market.

3.3 Software Tools

A new tool is being developed as part of the EPV Project to make the communication between Intelligent Charge Station Pod and the customer easy and feasible. Industrial partners with long experience in car industry are the leaders of this task, with the objective to obtain a strong, complete and functional system. Working with a Human Machine Interface (HMI), the final result is a graphic environment being pleasant to the user and be modular.

The next step on the development of control tools was to define a new architecture (see Fig. 4) to enable information exchange between charge station pods and the Control Centre software. The chosen technology needs to be based on standards to simplify integration and with special attention on the scalability of the communications solution and device protection certificate.

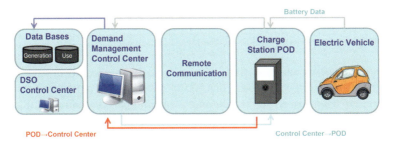

Fig. 4.   Communication architecture implemented at EPV Project

On the other hand, it is necessary to develop a software specification to manage the information over power grid and recharge infrastructure. It could be a demand management tool similar to any software developed to future smart grids. In the EPV Project architecture, two different supervisor and control centres are defined. A first version of this software links both applications in the same interface, so all the information can be easily reached. The first one is located on the utility side. This system is based in these parameters, which allows taking part on the power grid through real energy disposal knowledge and to decide, based on these data, when the batteries can be charged and what time is needed. Also, this system will provide a better integration of renewable energy resources like wind power by knowing the generation status in real time. This provides the equilibrium between power grid energy offer and charging pods demand.

The other software solution is located at the demand manager side which offers different tariffs adapted to the end user, applying internal methods for billing (see Fig. 5).

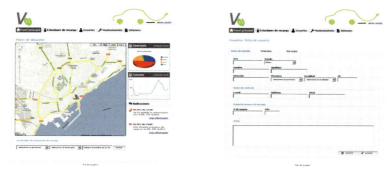

Fig. 5.   Demand Management Centre Software, Initial Screen and User Management Screen, EPV Project

This software is an important interface to change customer's awareness, of the environment effects of their consumption by the information regarding from where the energy is coming (that means particularly the use of renewable energies).

Taking into account the new trend of monitoring, control and supervisor software, the SOAP protocol is used, developing a webservice application to admit easy remote connection. The communications protocol between the utility and the infrastructure charge owner (business model) is the standard IEC 60870-5-104 (IEC 104) protocol, broadly introduced in Spanish utilities.

4   Conclusions

This paper is aimed at providing an overview of the most important aspects of the EPV project development, emphasizing communications aspects. A massive market penetration of electric vehicle would represent a change in the current society that involves more information and communication applications. Therefore, one of the main concerns of the EPV project is the standardization of the charging station pods, their characteristics and communications with the user, as well as the implementation of a management and billing system. A successful introduction of electrtic vehicles must be managed by software tools that allow utilities to integrate wind power as source to charge vehicle batteries, managing these intelligent devices as an active element inside the so called "Smart Grid".The new charge infrastructure needs to be studied, analyzed and optimized. From this project, we can get a great substantial benefit on the distribution network where a new business model for EV public charge has been detected. Additionally, the deployment of the required transport infrastructure would also be an important economic activity for the electrical network equipment manufacturers.

In the end, the EV is an active element inside the "smart grid"; how we can manage this intelligent device is a big concern for the EPV project. From an electrical perspective, the vehicle is a load for the power network but it could also be a generator too in certain conditions. Grid management complexity increases with this situation, but the EPV project will develop applicable models for this scenario. All these issues will make possible future studies for a better integration of electrically powered vehicles inside the Smart Grids.

References

[1] European Comission. Transport White Book. 2006.
[2] Abraham, A., et. al, Evolutionary Multiobjective optimization, Springer. 2005.
[3] AENOR. UNE-EN 61851-1 Sistema inductiva de carga para vehículos eléctricos. 2002.
[4] Mennekes Elektrotechnik GmbH & Co. PowerTOP Xtra. 2010.
[5] Saft Industrial Battery Group. Technical Manual DOC n° 21112-2-0704. 2004.
[6] Saft Industrial Battery Group. Batteries data sheet DOC n° 21532-2-0705. 2009.

Patricio Peral Galindo, Sixto Santonja Hernández
Energy Technological Institute
Juan de la Cierva – 24.
Parque Tecnológico de Valencia
Paterna
Spain
patricio.peral@ite.es
sixto.santonja@ite.es

Ana González Bordagaray
Iberdrola Distribución Eléctrica, S.A.
San Adrián – 48.
Bilbao
Spain
ana.gb@iberdrola.es

Keywords: wind energy, electric vehicle, charge infrastructure, ICT Tools, smart grid

Unifying Approach to Hybrid Control Software

M. Stolz, B. Knauder, P. Micek, W. Ebner, E. Korsunsky, Kompetenzzentrum
Das virtuelle Fahrzeug Forschungsgesellschaft mbH
P. Ebner, AVL List GmbH

Abstract

The presented work[6] focuses on a generic software architecture as
a basis for complex hybrid control and energy management strategies for a very wide range of applications. A clear and transparent
mode prioritization together with the requests generation form
the fundamental structure within the proposed architecture which
always ensures the secure handling of the complex system. Different
application dependent hybrid features can easily be attached onto
this core enabling calibration as well as testing in a straightforward
manner due to their capsulation. The lean interfaces between core
software and additional hybrid features are designed in a way to support software re-usability and scalability. Because of the very generic
approach, derived control software can serve the entire spectrum
from micro to full hybrid. The working principles of hybrid feature
selection, request generation and mode transition are presented in
detail for a selected example.

1 Introduction

Control and optimal energy management nowadays plays a key role in realizing the promised CO_2 reduction potential of hybrid cars by optimally utilizing
additional degrees of freedom given by them. Similar to conventional vehicle,
the hybrid vehicle shall satisfy the driver demands for different driving situations. Contrary to the conventional drive train, this can usually be performed
in hybrid vehicle with a number of different settings (hybrid operation modes)
for electric machine and combustion engine operation point (speed and torque
values), drive train configuration (transmission gear, state of clutches, engine
status: on/off), arrangement of energy flows, and state of auxiliary devices.
The primary task of hybrid vehicle control is to select at each instant the "optimum" settings of drive train components ("optimum" hybrid operation mode)
depending on the selected global operation strategy, actual driver demands,
driving situation and state of vehicle components, and to provide "smooth"
seamless transitions between the modes.

The definition of the "optimum" depends very much on the vehicle performance targets and the boundary conditions for their achievement. Usually, the goals aimed by hybrid vehicles include: 1) Improvement of the total efficiency of vehicle as compared to conventional vehicle. One of the consequences of this target is a reduction of fuel consumption. 2) Improvement or, at least, maintaining the same vehicle drivability and driving comfort as in conventional vehicles. 3) Reduction of emissions, noise and vibrations. At the same time, some important constraints shall be taken into account. For example, the availability and reliability of a hybrid vehicle and its components during at least pre-defined life time must be ensured.

There are several approaches to design the hybrid vehicle control. In general, one can divide the control strategies into optimal, sub-optimal and heuristic (or rule-based) ones [1, 2, 3]. The optimal strategies require detailed a priori knowledge about driving profile and include various optimization methods such as static optimization, dynamic programming, genetic algorithms, etc. The sub-optimal strategies apply these optimization methods and model-based driving condition prediction for a relatively short control horizon. The heuristic strategies are based on either deterministic or fuzzy rules derived using common relationships, expert knowledge, simulation models, and without a priori knowledge of driving cycles.

All of the strategies have their validity ranges with corresponding advantages and disadvantages. In some applications, for example for public transportation with fixed routes, the optimal strategies will obviously deliver the best result, while for passenger cars the sub-optimal or rule-based control methods shall be used. For practical implementation of the control software for mass production vehicles the aspect of software re-usability is very important. It is therefore desirable to develop such a software architecture that would enable implementation of different types of strategies or combination of them.

The re-usability of hybrid control software and its applicability in the scope of mass production hybrids imply, except for possibility to accommodate different control strategies, some other important requirements. The software shall be testable, also on a level of single software components, and be reasonably easy tunable (calibratable). The software architecture has to be able to deal with:
- ▶ Different topologies of parallel hybrid drive train.
- ▶ Different extent of hybrid functions and their implementation peculiarities.
- ▶ Different vehicle performance targets and operation constrains.
- ▶ Both energy optimization and drivability issues that require safe and robust handling of high-dynamic processes, for instance transitions between hybrid modes [4, 5].

The aim of the present work is to design architecture for the hybrid vehicle control software that would satisfy the above mentioned requirements.

In the following Section 2 the basic principles of the SW architecture, its structure and especially the idea of hybrid mode selection are described. Section 3 presents a particular example of the mode selection and dynamic mode transition to demonstrate the operation of the hybrid control software. Conclusions are drawn in Section 4.

2 The Generic Hybrid Control Software Architecture

2.1 Basic Principles of the Software Architecture Design

In general, the dynamics of the hybrid electric vehicle drive train belongs to the class of so called switched hybrid dynamical systems [5], which consists of continuous (quasi-stationary) phases interrupted by state jumps. The continuous phases include, for instance, pure electric driving, vehicle standstill with combustion engine switched off, recuperation, power assist (boost), etc. Each of these phases is characterized by a particular setting for the power split ratio between electric machine and combustion engine as well as set of parameters describing the drive train configuration, which includes the state of separation and launch clutches, the engine status and engaged transmission gear [1].

Therefore, in our approach to the hybrid vehicle control software, a suitably defined set of the quasi-stationary hybrid operation modes is a basis and starting point for the control concept. Generally speaking, this set of modes can be different for different hybrid vehicle variants. Exact definition of the modes is not a critical issue in this approach but it is important that the modes constitute a complete set covering the whole space of the vehicle operation points. For the sake of software re-usability it is however advantageous to define at least a basic set of "generic" hybrid modes that have common features for all hybrid variants, like recuperation, engine idle off, boost, pure electric driving, conventional operation, and so on.

Another important ingredient of the concept is the mode rating – a quantity for each mode which determines the "beneficence" and reflects the advantages of keeping this mode or switching into this mode.

All the hybrid modes are evaluated in parallel with respect to actual (and eventually, past and future) driving situation, driver demands, and state of vehicle components to calculate the relevant control parameters and the mode rating.

Depending on the rating, the "optimum" quasi-stationary hybrid mode with corresponding power split and drive train configuration requests is selected. These selected quasi-stationary requests are the inputs in the subsequent software component which treats the dynamic transitions between the modes and produces the final dynamic control signals.

This concept allows the software implementation of rule-based, optimal and sub-optimal control strategies for a wide range of different hybrid vehicle variants with a relatively small effort for adjustments in the software structure.

2.2 Software Structure

The basic layout of the hybrid control software architecture is shown in Fig. 1. The architecture consists of several layers responsible for the tasks described in the previous section.

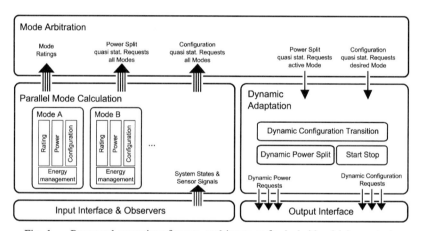

Fig. 1.   Proposed generic software architecture for hybrid vehicle control

The main objective of the first layer "Input Interface and Observers" is a collection and generation of data from hybrid components and vehicle necessary for hybrid control. Here, the signals received from sensors and other control units are filtered and pre-processed according to the requirements of hybrid control. Observers represent a "device abstraction layer": here the unified signals are produced for different drive train topologies, types of storage units, e-machines, combustion engines, etc. Device abstraction is one of the important issues allowing the implementation of control software functions in a generic way, such that the functions remain valid for very many different hybrid variants.

In a subsequent layer "Parallel mode calculation", the calculation of quasi-stationary requests and ratings is performed in parallel for each specified hybrid operation mode. The calculation is based on the used control strategy and, in general, includes evaluation of the actual, past and future driving situations, driver demands, and state of vehicle components. In particular, the state of energy storage units plays an important role in the definition of requests and ratings. It is therefore suggested to split the calculations up into four parts. The first one, "energy management" defines the ranges of available energy or energy capacity and power capability of storage units for each mode. These ranges are used then within the following three additional parts of mode calculation. Requests for mode specific torque split at demanded speeds as well as drive train configuration settings are generated within the second and third calculation step. Additionally a rating is calculated in a fourth calculation step. The mode rating calculation algorithms depend on the vehicle performance targets, may include various constrains (e.g. how often the battery may be loaded). They also may be different for different driving situations to reflect optimization of specific targets (e.g., at vehicle launch the targets are to support the desired lifetime of drive train components, reduce noise and emissions; at driving with constant velocity the target maybe to reduce the fuel consumption).

The algorithms for mode requests and ratings may range from simple rule-based estimations using pre-calibrated maps up to dedicated optimization algorithms. The proposed here splitting of mode calculation into four parts is not necessary, but might be useful considering the aspects of software re-usability and testability.

The mode arbitration and selection, based on the mode rating comparison as well as pre-defined mode prioritization, takes place in the software layer "Mode arbitration". Two groups of control parameters for hybrid electric vehicle can be identified with respect to their dynamics: 1) "operation mode configuration" parameters that include the drive train configuration parameters and state of auxiliaries; 2) "fast quantities" like engine and e-machine torque/speed. The operation mode configuration can be considered as a frame for parameters of the second group: correct configuration is needed to implement the torque and speed requests. The implementation of requests for the "fast quantities" proceeds in normal operation relatively quick and safe. In contrast, the implementation of requests for operation mode configuration proceeds usually on much longer time scales or may even not be realized at all. Therefore, two modes are selected in our approach: the first one corresponds to the actual absolute highest rating, the second one – to the highest rating only for modes with actual configuration. Thus, the operation mode configuration is requested for the first selected mode (best possible mode), while the torques and/or speeds are requested for this mode only when the configuration requests are implemented (actual configuration = requested configuration). Until then only those requests are issued that don't need re-configuration.

The dynamic transitions between the modes are treated separately in the software layer "Dynamic adaptation". The dynamic adaptation of the requests involves coordination of the drive train configuration changes such as activating/deactivating the clutches, gear shift, starting or stopping the combustion engine, as well as the necessary for transition changes in torque and speeds, like synchronization of engine and electric motor speeds. This layer may be split into three respective parts to provide software modularity and testability.

The architecture presented here provides a basis for real scalability of the hybrid control software. The configuration and power split requests for each operation mode are defined within the mode calculation itself, while mode transitions are not covered within a mode and treated separately. As a consequence the mode calculations can be kept rather independent from each other. This allows adding or replacing the operation modes, if needed for different hybrid vehicle variant, without adjustment of the other modes, without changes in the whole structure, and with only small adaptations in the mode selection algorithm.

3   Example of Hybrid Modes and Mode Transition Control

In this section, the functionality of the control software built using the described above basic principles and architecture is demonstrated on the example of few modes and transitions between them during the operation of test hybrid vehicle. The drive train of the test vehicle with parallel hybrid topology includes the gasoline engine which can be decoupled from the electric machine by automatic hydraulic clutch. Manual transmission with a manual launch clutch connects these two power sources to the rear wheels.

Fig. 2 shows three different hybrid operation modes and transitions between them at vehicle standstill (vehicle velocity is ~0km/h and the launch clutch is open, separation clutch is closed). The first mode (denoted in Fig. 2 as phase 1) is the conventional mode where the combustion engine works in idle. The second mode (phase 3) is the "engine-idle-off" mode where the combustion engine is stopped. This mode is in general used to save the fuel if energy amount in the storage system is sufficient to supply the electrical board net and auxiliaries. The third mode (phase 6) is the "idle charge" mode where the combustion engine works in idle and drives the electric machine as generator to produce electrical energy to charge the storage system. The phases 2, 4 and 5 are transitions between the modes where the drive train configuration is changed.

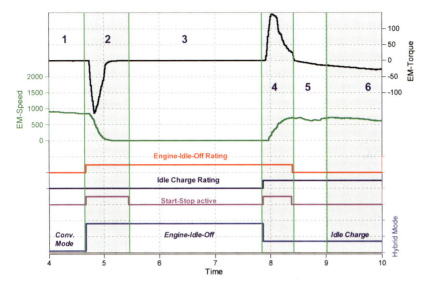

Fig. 2.   Measured signals and control parameters for quasi-stationary modes "conventional" (phase 1), "engine-idle-off" (phase 3), "idle charge" (phase 6) and transitions between them (phases 2, 4, 5) at operation of parallel hybrid test vehicle.

In the phase 1, the vehicle is in conventional mode which has the highest rating (not shown in Fig. 2). In this case the engine is running in idle. In the second phase the rating of the engine-idle-off mode is increasing because some specific activation condition for this mode (e.g. duration of engine idle exceeds pre-defined threshold) gets fulfilled, and the rating gets higher than the conventional mode rating. The "Mode Arbitration" module (see Fig. 1) checks that the engine-idle-off rating is the highest one, selects engine-idle-off as the best possible mode with its corresponding drive train configuration request. In this particular case, the combustion engine is requested to be switched off. This request is forwarded to the software layer "Dynamic adaptation", where the Start-Stop module decides (based on information from observers) to perform "active" engine stop with pre-defined engine speed ramp and issues corresponding request for electric machine speed (remark: since the separation clutch is closed, the speeds of the engine and e-machine are equal, see Fig. 2). This dynamic speed request is transformed into the torque request inside the module "Dynamic power split", where it is also limited in gradient and finally sent via CAN to the electric machine control unit. As can be seen, the e-machine torque and speed during the transition phase (phase 2 in Fig. 2) are completely under the control of the software part "Dynamic adaptation". As soon as the transition is finished (indicated by setting down the flag 'Start-Stop active' at the end of phase 2 in Fig. 2), the power split request from actual quasi-stationary mode is used to produce final commands for the electric machine and the engine. In the case of engine-idle-off (phase 3), both engine and e-machine torque requests are zero.

In a similar way the transition to the next hybrid operation mode takes place. At the beginning of phase 4, the rating of the idle charge mode increases since the battery state-of-charge falls below the pre-defined threshold. The idle charge rating gets equal to that of engine-idle-off mode. However, since the priority of idle charge is higher, the "Mode Arbitration" (where priorities are defined) selects the idle charge mode with its corresponding drive train configuration request. This time it is the request for engine running in idle, which goes to the Start-Stop module. The Start-Stop module decides to perform engine speed-controlled start with pre-defined engine speed ramp. As for the phase 2, this speed request is transformed into the torque request inside the module "Dynamic power split", where it is also limited in gradient and sent via CAN to the e-machine control unit. When the engine reaches idle speed (indicated by setting down the flag 'Start-Stop active' at the end of phase 4 in Fig. 2), the power split request for quasi-stationary mode idle charge gets active. This request however implies negative torque of the e-machine. It would have unacceptable influence on the engine speed (and may even lead to stall of the engine) when switched immediately after the phase 4. Therefore, the software module "Dynamic power split" filters this negative torque request to provide a smooth torque transition (phase 5). The phase 6 represents a quasi-stationary mode idle charge.

4   Conclusions

The generic control software architecture presented above provides an easily scalable function framework for a wide range of hybrid vehicle topologies and their various applications (engineering targets settings), which has also been successfully verified by several applications e.g. in a fully hybridized passenger car. The correct functionality was proved by presented measurements.

The core concept of the presented paradigm is the decomposition of the requested functionality into a set of encapsulated operating modes, which are computed in parallel. The selection of the actual operating mode takes into account its actual "level of optimality" (with respect to concurrent modes) and current drivetrain configuration.

The presented concept shows high potential for effective hybrid control software development and its calibration under various target settings and for a wide range of hybrid topologies.

5   Acknowledgements

The authors would like to acknowledge the financial support of the "COMET K2 - Competence Centers for Excellent Technologies Programme" of the Austrian Federal Ministry for Transport, Innovation and Technology (BMVIT), the Austrian Federal Ministry of Economy, Family and Youth (BMWFJ), the Austrian Research Promotion Agency (FFG), the Province of Styria and the Styrian Business Promotion Agency (SFG).

We would furthermore like to express our thanks to our supporting industrial and scientific project partners, namely AVL List GmbH. and to the Graz University of Technology.

References

[1]   Guzella, L., Sciarretta, A., Vehicle Propulsion Systems. Introduction to Modeling and Optimization, Springer-Verlag, Berlin Heidelberg, 2007.

[2]   Salmasi, F., Control strategies for hybrid electric vehicles: evolution, classification, comparison and future trends, IEEE Transactions on vehicular technology, Vol. 56, 2393, 2007.

[3]   Wirasingha, S., Emadi, A., Classification and review of control strategies for plug-in hybrid electric vehicles, IEEE Transactions on vehicular technology, Vol. 60, 111, 2011.

[4]   Beck, R., Saenger, S., Richert, F., Bollig, A., Neiß, K., Scholt, T., Noreikat, K.-E., Abel, D., Model predictive control of a parallel hybrid vehicle drivetrain, Proc. 44th IEEE Conference on Decision and Control – European Control Conference, Seville, 2670, 2005.

[5]   Koprubasi, K., Westervelt, E., Rizzoni, G., Toward the systematic design of controllers for smooth hybrid electric vehicle mode changes, Proc. of the 2007 American Control Conference, New York, 2985, 2007.

[6]   This work was supported in part by COMET K2 - Competence Centres for Excellent Technologies Programme.

Michael Stolz, Bernhard Knauder, Petr Micek, Wolfgang Ebner, Evgeny Korsunsky
Kompetenzzentrum Das virtuelle Fahrzeug Forschungsgesellschaft mbH
Inffeldgasse 21/A/I
8020 Graz
Austria
michael.stolz@v2c2.at
bernhard.knauder@v2c2.at
petr.micek@v2c2.at
wolfgang.ebner@v2c2.at
evgeny.korsunsky@v2c2.at

Peter Ebner
AVL List GmbH
Hans-List-Platz 1
8020 Graz
Austria
peter.ebner@avl.com

Keywords: HEV, mode transitions, mode arbitration, modular software, software framework, supervisory control, driveability, hybrid vehicle control strategy, smart management and routing of energy and power

Smart Power Li-Ion Battery Systems for the Safe and Networked Mobility Society: The ICT Potential

R. John, Infineon Technologies AG
T. Schäfer, Li-Tec Battery GmbH

Abstract

HEV-EV batteries are characterized by high specific power, energy, efficiency and long life. These batteries will soon play a prominent role as innovative electrochemical storage systems in renewable energy plants and distributed power stations as well as in smart power systems for sustainable vehicles such as hybrid and electric vehicles. There is now a new dimension of improved safety, reduced costs, greater operational temperatures and materials availability while innovations based on ICT are set to be an important mainstream element in the near future.

1   The New Challenge for Smart Power - 8 Requirements

In Fig. 1, a total of eight requirements are defined for EV (dark line) and HEV (red line) which differ significantly from those of other applications, and all which must be exceeded for (H)EV applications. Particular cases in point for the so-called Consumer CCC market segment are Cold Crank Power, the total lifetime, as well as the cost facing battery system developers. In terms of the development trends of recent years, it can be stated that, generally speaking, energy density and capacity have increased, as has reliability, and they continue to be developed. From the sustainable criteria standpoint, reduced use of toxic materials such as cobalt and nickel can accompany embracing manganates or the currently widely-discussed iron phosphates.

It is widely accepted that the ionic battery family, in particular the Li-Ion systems, offers a wide spectrum of system approaches which have been optimized to meet these requirements or are able to be further evolved in their development. In addition, the battery systems have a certain revolutionary potential for multiplying volumetric energy and reducing costs with longer-lasting ionic battery systems just within the next few coming years alone. As a developer of batteries and in the interest of the entire battery industry as well as automobile manufacturers, the author points out that special attention must be paid particularly to material and environmental policy issues if we want to consider and develop European locations and sensibilities.

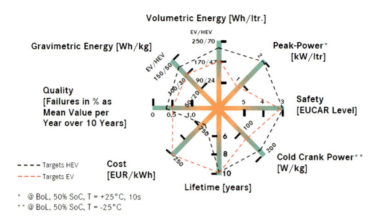

Fig. 1.   8 Requirements need to be fullfilled for (H)EV batteries; Source: Daimler AG, 2011

Ionic energy storage devices require deeper and more specific focus on production and quality control in light of addressing completely new challenges, for example in Germany with production research programs such as PROLIEMO and DELIZ (KOPA II). It was recognized that new approaches needed to be adopted in order to scientifically realize the strategic goals and also develop the necessary knowledge. In terms of safety through to ultimate disposal, the new energy storage devices can only be produced and designed to comply with the EUCAR levels. In this respect, new approaches to production testing systems might need to be established for the entire life cycle of such batteries (external control, accompanying reliability tests); the international Battery Safety Organization project (BATSO), Taiwan, being one such example.

Yet it may well be pointed out that when one globally analyzes the value chain, there is a lack of R&D integration and inclusion in essential industrial activities right here in Germany. This should by no means be underestimated considering the cluster structures in America and Asia. In this respect, the author suggested an R&D center and battery quality and reliability testing center at the Dresden University of Technology for the activities of the National Platform for Electric Mobility.

The following will take a brief look at questions, ideas and approaches to using ionic batteries, including their integration and management, within the scope of ICT and Smart Power solutions.

Recent technological developments in batteries and battery assemblies, for example those seen in new current interrupt devices (CID), Protection One Chips (POC), smart protection modules for cells and assemblies, Battery Management Systems (BMS) and Battery Management Monitoring Systems (BMMS) in combination with advanced ICT and innovative ideas will lead the new EV age toward a new quality of individual green e-mobility.

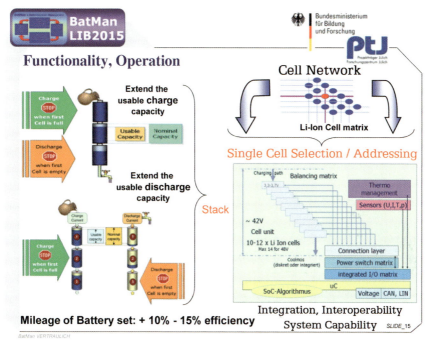

Fig. 2.   BATMAN Project slide; Source: Reiner John, Infineon, 2011

The battery management system, for instance, is on the one hand required to ensure the lifetime of the battery, which it does by making sure that no cell in the battery functions outside of its operational range; at the same time, it's a required safeguard ensuring the battery is not operated outside of safe limits. In multi-cell or module configurations, there are additional requirements, the basics of which will be touched on briefly. The Smart Power Li-Ion system of the future will likely provide other new additional features useful to the user such as very accurate State of Charge (SOC) indications. It will also enable the user to assess the battery system's State of Health (SOH).

2   The BATMAN Project and Smart BMS Solutions

The battery resulty from the Li-Tec Battery project (JV Evonik & Daimler) is thus also a part of the BMBF LiB 2015 project initiative in the joint BATMAN project, with the continuing instrumental participation of Bosch and Infineon.

BATMAN does not refer to bat-like creatures but rather the project title for "Innovative battery sets based on modular high-performance Li cells and modular BMS for use in harsh environments," hence whisper-quiet, low-emission regional driving or efficient, safe storage of somewhat regenerative electricity, thereby enabling the recharging of electric cars or the establishing of decentralized power supplies.

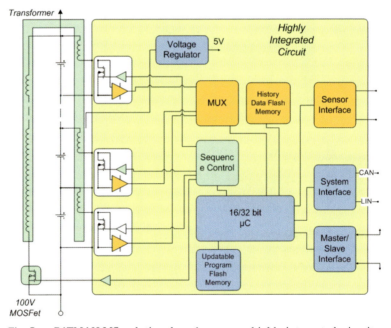

Fig. 3.   BATMAN-MS solution based on new highly-integrated circuits; Source: Infineon

In concrete terms, the BATMAN project is researching new systems for electronically-supported management of such lithium-ion batteries in hardware, design and application. Wide applicability from e.g. light electric vehicles to electric cars is one goal. On the other hand, there is also the issue that expected to be particularly safe and reliable. Their battery management electronics should additionally be reliable for 10 years and allow maximum failure rates of 10 ppm. This should also ensure an operating window for reliable BMS electrical function down to -40° C. The goal and the challenge of the R&D work lies in demonstrating the technical capacity of using such new system integration concepts based on new control chips and algorithms.

There is evidence here that the R&D project has succeeded in resolving the paradox between minimizing the number of circuits and their complexity (costs, reliability) and the contrasting demand for higher flexibility. In the project, Bosch is focused on a battery set for electromobility, Gemac and Clean Mobile on light electric vehicles, and Li-Tec Battery is applying project findings to the "Lessy" large-scale storage device project as well as a project being carried out within the scope of the German LiB 2015 BMBF initiative.

All the partners have integrated basic Infineon components such as Power ICs, current sensors and balancers. At the end of the project, the partners intend to continue work on the projects or realize them in applications as soon as possible.

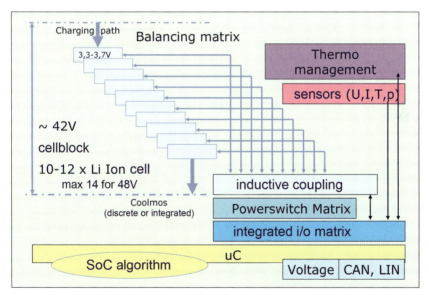

Fig. 4.   Example: Electronic Functions for 42 V Cell Stack, BATMAN Project

The illustration below depicts the basis and functionality of the BATMAN BMS using an example 42 V solution with fundamental functions based on the new IC solution.

Integrable sensors or completely new mechanisms or operational concepts are seen as being able to be extended with this BATMAN approach to Smart Power concepts. New safety precautions should already be incorporated hereto in the near future. Ultimately, our entire society and individual mobility benefits from realizing and innovating completely new solutions this way. The following will outline part of an basic example of this. There are currently different operational concepts related to e.g. replacing or recharging batteries. In this respect as well, ICT-based Smart Power approaches provide particularly novel procedures for more comfort, efficiency, reliable supply, vehicle operation, navigation, vehicle energy control systems, supply stations, etc.

3   ICT-Based Navigation Systems and Smart Power

Navigation systems storing and using information about fuelling station status are already known. Yet this information maybe alone is insufficient for electric drives. Due to the lower energy density of electric power units (batteries) compared to conventional fuels, and the correspondingly increased need for storage volumes, as well as due to the potential variety of batteries, a sufficient reliable supply requires greater logistics than is the case with conventional fuels. Hence, operation must be based on a mix of recharging, redistributing between supply stations, and local charging of batteries under flexible consideration of ongoing fully-charged outflow versus inflow of empty or partially empty batteries at the supply stations. Should no fully-loaded batteries of the respectively suitable type be available at the time a vehicle driver needs a charge, or even more importantly a replacement, the driver is left with the time-consuming charging and thus forced to stay at the supply station.

This is especially unpleasant at fully automatic stations without lodging options, particularly in otherwise sparsely populated and climatically unfavorable areas. On bodies of water, in particular larger lakes or offshore waters, the above-described concern takes on special significance in terms of safety. If a boat or a ship is electrically operated on larger bodies of water, for instance Lake Constance (Germany) or other lakes, it then becomes important to always remain within range of a battery charging or changing station to avoid unexpectedly ending up in distress. On the other hand, it is aggravating to always be excessively cautious in staying within the vicinity of supply stations when not absolutely imperative. ICT can help solve such exemplary or basic smart power solutions and to reduce time to market.

References

[1]   John, R, Batterie Management für mobile Lithium-Ionen Energiespeicher BatMan - Innovatives Batteriemanagement und Architekturen für Energiebereitstellung, Verteilung, sicheren Betrieb und hohe Lebensdauer von mobilen Lithium-Ionen Batterien - , proceedings: Innovationsallianz LIB 2015, Westfälische Wilhelms-Universität Münster - 2010.

[2]   Schäfer, T, Verfahren zum Betreiben eines Fahrzeugs, PCT/EP 2010 002070.

Reiner John
Infineon AG
Am Campeon 1-12
85579 Neubiberg
Germany
reiner.john@infineon.com

Tim Schäfer
Li-Tec Battery GmbH
Am Wiesengrund 7
01917 Kamenz
Germany
tim.schaefer@li-tec.de

Keywords: Smart Power, Li-ion Batteries, safety BMS, sensors, HEV-EV, renewable energy, grids, requirements, LIB 2015, BATMAN, BMBF, infineon, bosch, Li-Tec Battery

Safety & Driver Assistance

Low-Cost Platform Technology for LWIR Sensor Arrays for Use in Automotive Night Vision and Other Applications

I. Herrmann, M. Hattaß, D. Oshinubi, T. Pirk, C. Rettig, K. F. Reinhart, E. Sommer, Robert Bosch GmbH

Abstract

In the EU FP7 project "ADOSE" a new approach has been used to develop a new cost efficient technology by adapting a volume proven integrated MEMS process for the production of a suspended thermodiode array. As low-cost was to be the key feature of the project the focus was set on fully semiconductor compatible production without the need for dedicated equipment and on the use of cost efficient MEMS technologies like wafer-level vacuum encapsulation. A first proof-of-concept integrated FIR array with 42 x 28 pixels was already published [1], now the final ADOSE design with 100 x 50 pixel resolution is produced and under evaluation. While the ADOSE chip targets the requirements of a low-end "Hot-Spot-Detector", many applications require higher resolutions. As the technology used in ADOSE inherits some limitations from its pressure sensor ancestor, we advance the technology to get rid of these and achieve a low-cost and high-res imager. Here we present the results from ADOSE and give an outlook on the new BMBF-funded Spitzencluster MicroTEC Südwest project RTFIR.

1   Introduction

The EU FP7 project ADOSE targets a variety of low-cost and high performance sensor technologies that enable reliable detection and classification of vulnerable road users and obstacles and through this enhance active safety functions. The project is focused mainly on the sensor technologies, the sensing elements and some pre-processing hardware and only to a minor degree on system concepts and software. Besides the thermal infrared camera presented in this paper the following sensory concepts are part of ADOSE: multifunctional CMOS imagers, 3D packaging technologies, ranging techniques, bio-inspired silicon retina sensors, harmonic microwave radar and tags.

The Spitzencluster MicroTEC Südwest project RTFIR is focused on thermal imagers but targets also applications abroad automotive like industrial process control, ambient assisted living and security technology. As in ADOSE the projects deals with the technological aspects and the sensing principle primarily and not with the system or software aspects other than needed for testing and demonstration of the technology.

1.1 Automotive Night Vision

Currently available high-end systems in luxury and upper class cars either focus on excellent image display and use the combination of a NIR CMOS imager and additional NIR headlights or make use of stand alone FIR night vision with expensive high resolution bolometers. Next generation automotive night vision systems for driver assistance will improve the safety of vulnerable road users with active warning signals and in future systems also automatic system action. Reliable detection with low false alarm rates is essential for such systems. With current technology only the combination of NIR and FIR is able to fulfil all requirements but this makes the systems even more expensive. A fused NIR/FIR system combining high resolution NIR and lower resolution FIR would allow excellent image display quality using the active NIR image from affordable good resolution CMOS image sensors and at the same time provide the additional information required to reliably identify vulnerable road users through the support by hot spot detection from a low-cost mid-res FIR-array sensitive in the 7-14 µm wavelength range (see Fig. 1).

Hot-spot detection has some advantages towards low-cost realization compared to stand alone FIR imaging as the resolution requirements both in thermal and spatial view are significantly lower when the thermal imager is combined with a VIS/NIR CMOS camera with higher spatial resolution. The latter can then be used for purposes like road sign detection, lane departure warning or others using image recognition anyway while the thermal imager provides the information about thermal properties of objects.

2 Requirements

2.1 ADOSE

In the specification part of ADOSE different use cases were analyzed and the use case of extra-urban single lane roads with a maximum velocity of 100 km/h was chosen as the base for the specification of the sensor modules. For a safe break in this scenario a detection distance of 120 m was calculated. The minimum hot spot for detecting a human being has been set to 1 by 5 pixels (hor. x ver.). Together with the viewing angle necessary for usual curved roads of $\pm 12°$ the spatial resolution of 100 by 50 pixels was derived. The thermal resolution was specified to 0.5 K.

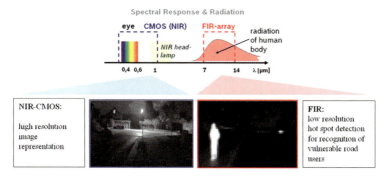

Fig. 1.   Multi-spectral approach for warning night vision with NIR / FIR data fusion (top); identical night vision scene with different sensors (bottom): CMOS-NIR imager without active light (left) and low resolution FIR-array (right).

Looking at the cost for the imager chip used in an FIR add-on module only a few tens of Euros seem to be acceptable, because there is another major cost part, the optics. We looked into low-cost FIR optics in ADOSE as well but the results from this will be published later.

FIR camera requirements		*Remark*
Hor. Field of View (FOV):	± 12°	For data fusion with NIR
Angular Resolution:	4,18 pixel / °	Defined by smallest object to be resolved @ 120m
Object Temperature resolution:	< 500 mK	for hot-spot detection; no greyscale image display NETD < 300mK for chip @ F#1 optics
Frame Response:	> 12,5 Hz	for 3 verifications of object in the NIR image
Array Size:	100 x 50 pixels	Defined by FOV and angular resolution
Wavelength Range:	7-14 µm	Spectral emission maximum of vulnerable road users

Tab. 1.   ADOSE FIR add-on module specifications

2.2 RTFIR

In RTFIR the required specification comes form several different application with different requirements. For the automotive application the target is to address not only the limited use case of ADOSE but also use-cases were higher spatial resolution are necessary. To achieve this, the sensor technology and the read out concept will have to focus on scalability so a wide range of resolutions may be produced just by altering layout to utilize economy of scales. This effect would vanish when different resolution would require different processing and therefore the "low-cost" would be feasible for high volume applications only. Because of limited resources the somewhat standard resolution for un-cooled thermal imagers of 320 x 240 has been chosen as the single resolution to proof and demonstrate the concept and technology. For the thermal resolution the target is 100 mK for the automotive set-up.

3 Sensor Technology

3.1 ADOSE

As the technological concept of ADOSE was already described in detail in [1], here we just present a short overview of the process and concentrate on the design improvements made since then.

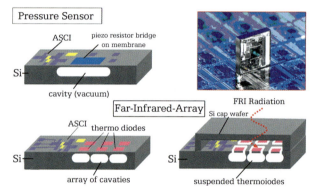

Fig. 2.   From integrated pressure sensor to thermal imager. Top left: integrated pressure sensor principle; top right: picture of an integrated pressure sensor, the membrane is the square in the centre of the chip; bottom left: Instead of a single cavity for the pressure sensor an array of cavities is made (one beneath each pixel); bottom right: fully vacuum packaged thermal imager.

In order to make use of the existing mass production and cost effective RB pressure sensor technology we decided to differ from the standard resistive sensing principle of microbolometers and use the forward voltage of a diode made in monocrystalline silicon. This type of sensing element is available in the IC process of the integrated pressure sensor without the need of additional masks or material layers and comprises very low noise. Also we used the dielectric layers of the process as the absorption structure to save costs for a reflector and λ/4-structure as are used in microbolometers. The thermal insulation of the sensing element is achieved by structuring the pressure sensor membrane. The vacuum encapsulation is done by a silicon cap bonded on wafer level using glass-frit bonding as it is known from inertia sensors (see Fig. 2).

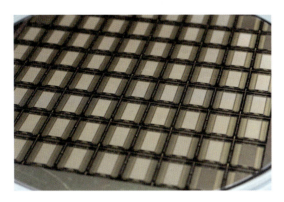

Fig. 3.   Photograph of first ADOSE wafer

3.2. RTFIR

The process for RTFIR is not finally chosen, but the concept is clear. The main inhibitors for higher resolutions within the ADOSE process are the restriction of process parameters coming from the combination of a MEMS and an IC process. So the first step is to separate the MEMS and the ROIC and the second step is to find a way to contact both as a classic two chip approach is not feasible with a large pixel array because of too many connections and the vertical integration approach of microbolometers where a CMOS ROIC is used as the substrate for the MEMS layers is too expensive. With the separation of the two parts both of them may be optimized independently, e.g. the membrane thickness for the MEMS may be reduced and the thermal regime of the process steps has not to be limited to CMOS compatibility levels. For the ROIC modern CMOS processes with 0.35 µm structures or lower may be used to increase functionality and decrease chip size and cost as well.

4   Pixel Design

The main issue for the design of the sensor pixel is to maximise thermal insulation and fill factor to get the highest signal out of the radiation incoming from the object observed. The second issue is to keep the thermal capacity of he pixel sensing element low in order to avoid high thermal time constants that would lead to smearing in the image.

In our most recent AMAA publication [1] we showed the thermal design and simulation of such a pixel in the ADOSE process. Unfortunately this design did result in too little signal and even if the insulation could be increased sufficiently, this would lead to too high thermal time constants, so we had to improve both parameters. One of the advantages of our forward current biased diode concept is the possibility to use more of them in a row to scale the signal. As the intrinsic noise of the monocrystalline diodes is very low, the SNR just scales with the number of diodes in series. Therefore we decided to place four diodes on the pixel structure instead of just one. A side effect of this was to be able to reduce the thermal mass by etching between the diode subpixels. Another measure taken to increase the signal was to improve the thermal insulation by using a different material for the electric contacts of the diodes. Fig. 4 shows the simulation results for such a four-fold diode pixel.

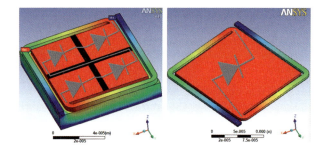

Fig. 4. Thermal simulation of a four-fold diode ADOSE pixel compared to the old single-diode design

For the RTFIR pixel design the main issues are a massive shrink of the pixels to keep the chip for the larger arrays at reasonable size and a drastic increase in insulation to countermeasure the loss of absorbed power resulting from the decreased pixel size. Table 2 shows the design properties of our ADOSE pixel and gives an outlook on the planned properties of the RTFIR pixels.

	ADOSE	RTFIR
Pixel Pitch [μm]	100	25
Absorber area [μm²]	5000	256
Fill factor [%]	47	44

Tab. 2. Design properties of ADOSE and RTFIR pixel

5   Read-Out Concept

To achieve the frame-rates specified in ADOSE the ROIC was designed with a massive parallel pre-amplification stage. This stage was built as a switched-capacitor amplifier unit which allows adjusting the pre-amplification level and therefore the temperature range of the sensor by external timing. The parallelization was made column-wise, so one row is addressed and in each column the active pixel in that row is differenced to the reference pixel of that column and the difference is amplified by the SC-integrator. The analog output amplifier stage comprises an offset correction feedback, so fixed-pattern noise can be eliminated before the final amplification. Fig. 5 shows a schematic view of one pre-amplifier stage.

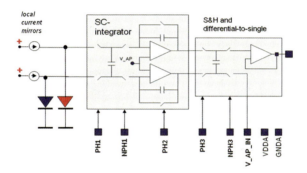

Fig. 5.   Block diagram of SC integrator stage.

6   Results

First wafers with the ADOSE pixel and ROIC design have been produced and are currently under test. The thermal responsivity of the pixels meets the simulated values of about 8000 V/W, while the thermal time constant exceeds the simulated value. The reason is still under investigation.

7   Discussion and Outlook

The improvements to the pixels we applied seem to work as expected so after completing the implementation of the chip into the demonstrator board we expect to achieve most of the target specifications of ADOSE. The issues left are the thermal time constants and some cross-talk problems in the ROIC. We expect to be able to complete a functional demonstrator before the end of the ADOSE project which is planned for August 2011.

In the next generation low-cost thermal imager project RTFIR we expect to improve the performance and specifications of our imagers significantly without adding major costs.

8   Summary

A new approach to low-cost thermal imagers has been proposed and proven with a functional 42 by 28 pixel array. The improvements necessary to meet the ADOSE requirements were worked out and first silicon containing the according thermosensor-array has been produced. Also the concept for a future platform technology for low-cost thermal imagers with a wider resolution range has been described that may enable the application of thermal imaging at much lower cost and therefore bring the benefit of life-saving night vision system to middle and even compact class cars.

A part of the research leading to these results has received funding from the European Community's Seventh Framework Programme under grant agreement n° 216049, relating to the project 'Reliable Application Specific Detection of Road Users with On-Board sensors (ADOSE)'. Another part of this work was done within the Spitzencluster MicroTec Südwest funded by the Bundesministerium für Bildung und Forschung.

References

[1]   Reinhart, K.F., et al., Low-cost approach for Far-Infrared Sensor Arrays for Hot-Spot Detection in Automotive Night Vision Systems: In: G. Meyer, J. Valldorf and W. Gessner, Advanced Microsystems for Automotive Applications 2009, Springer Berlin Heidelberg, 2009.

I. Herrmann, M. Hattass , D. Oshinubi, T. Pirk, C. Rettig, K.-F. Reinhart, E. Sommer
Robert Bosch GmbH
Corporate Sector Research and Advance Engineering
Microsystem Technologies, CR/ARY
P.O Box 10 60 50
70049 Stuttgart
GERMANY
ingo.herrmann@de.bosch.com
mirko.hattass@de.bosch.com
dayo.oshinubi@de.bosch.com
tjalf.pirk@de.bosch.com
christian.rettig@de.bosch.com
karl-franz.reinhart@de.bosch.com
edda.dommer@de.bosch.com

Keywords: thermal imaging, night vision, vulnerable road user, sensor array, diode, low-cost, ADOSE, RTFIR, MicroTEC Südwest

Generalized Space Frequency Representation Techniques for Enhancement and Detection Under Low Light Conditions

P. Kharade, S. Gindi, V. G. Vaidya, KPIT Cummins Infosystems Ltd.

Abstract

Night vision systems are becoming more useful as driver assistance systems as the number of accidents during night time is higher than in day time. The images taken in low light conditions are very poor in quality; the pedestrians and animals are hardly visible due to the darkness. In this paper, we have presented two different night vision applications using Generalized Space Frequency Representation (GSFR). The first part of the paper describes a method for image enhancement using GSFR with an exponential kernel. The enhancement includes improvement of brightness, removal of noise and improving the sharpness. The GSFR helps to remove noise from image, brightness and contrast is improved due to the use of the exponential kernel and finally sharpness improvement is done by using unsharp masking based on partial derivative. The second part of the paper describes a method for pedestrian detection using GSFR. It includes calculating weighted average pixel value based on pixel's orientation, applying modified pear shaped curve, applying GSFR, dividing image into layers and then finding a pedestrian in these layers using connected component labelling.

1 Introduction

Night Vision systems are being developed rapidly to reduce road accidents. These systems help the driver to get a clear view of the scene ahead of the vehicle, detect obstacles on the road and warn driver accordingly. We present an image enhancement and pedestrian detection methodology to use in night vision systems. The technique helps overcome poor visibility problems in an image due to fog, snow, and rain. The algorithm processes noisy images and gives better visibility to driver to identify obstacles. It also detects pedestrians and warns driver. Night vision system involves capturing image with NIR (Near Infra red) CCD camera and enhancing the image with proposed scheme. Enhanced image can be displayed to the driver and the driver can be alerted, using audio-visual alarms indicating the presence of pedestrians.

This technique involves two modules: image enhancement and pedestrian detection. The description in this paper begins with a discussion about the prior work in this area. The theory of GSFR and its associated parameters are explained in the next section. How GSFR can be used for the image enhancement, is explained in the subsection; followed by explaining the methodology involved in this pedestrian detection algorithm. Experimental results are reported that evaluate the performance of this technique in a night vision system. Accuracy is described in terms of percentage correct detection of the algorithm. Performance characterization is done against noise and contrast. Finally, the paper is concluded with concluding remarks, comments and claim of robustness that can be achieved by our algorithm. Future work is also briefly discussed in the conclusion.

2   Related Work

Many researchers have been working in the areas of night image enhancement, super-resolution image, multi-spectral imaging, night vision goggles and image fusion, to combat night vision issues. In image enhancement, histogram equalization method is commonly used. However, histogram equalization considers global image information to process the image, resulting in improper enhancement at object or local level [1]. In super-resolution or multi-spectral techniques, one more imaging modality is required. This increases overheads on the hardware. Additionally, speed is a major concern in such applications and thus it cannot be compromised. Night vision goggles suffer from disorientation and offer a small field of view to the user [2]. Many researchers have also focused on pedestrian detection. Effective pedestrian detection has been promised using FIR cameras [3],.The disadvantage of using FIR camera is that there are a number of factors that affect temperature in outdoor scenes hence the scenes viewed are difficult to process.

In our approach a Near Infrared Camera is used to capture the video ahead of the vehicle. It is a simple CMOS camera with IR sources which costs less than the FIR camera. Image enhancement helps make a view clear. Also further processing doesn't get affected by environmental factors like fog, rain etc. We present a robust algorithm for pedestrian detection with the detection accuracy of 63.3% even in the scenarios like dark area.

3   Methodology

3.1   Generalized Space Frequency Representation

Wigner distribution is a special case of the Generalized Space–Frequency Representation (GSFR). The characteristic of the Wigner distribution to be a function of both time and frequency is remarkable. The Fourier transform on the other hand, is strictly a function of frequency [4]. Wigner [5], in 1932, proposed this function for the study of quantum mechanics. Ville [6] proposed it again in 1948. However, researchers did not pay much attention to the method until the 1980s, when researchers in the speech processing area, extensively used this concept in 1980s.

The Wigner distribution of two signals $f(t)$ and $g(t)$, is defined as

$$WD_{f,g}(t,\omega) = \int_{-\infty}^{\infty} e^{-j\omega k} f(t+k/2) g^*(t-k/2)dk \tag{1}$$

Where ω is the frequency, t is time and g^* is the complex conjugate of the function $g(t)$.

The auto-Wigner distribution of signal $f(t)$ is given by

$$WD_f(t,\omega) = \int_{-\infty}^{\infty} e^{-j\omega k} f(t+k/2) f^*(t-k/2)dk \tag{2}$$

The auto-Wigner distribution for real function $f(t)$ is given by

$$WD_f(t,\omega) = \int_{-\infty}^{\infty} e^{-j\omega k} f(t+k/2) f(t-k/2)dk \tag{3}$$

In discrete domain the Wigner function is defined as

$$WD_{f,g}(t,\omega) = 2 \sum_{k=-\infty}^{\infty} e^{-2j\omega k} f(t+k) g^*(t-k) \tag{4}$$

The auto-Wigner distribution is defined as

$$WD_f(t,\omega) = 2\sum_{k=-\infty}^{\infty} e^{-2j\omega k} f(t+k)f^*(t-k) \qquad (5)$$

If the function is real then auto-Wigner distribution is defined as

$$WD_{f,g}(t,\omega) = 2\sum_{k=-\infty}^{\infty} e^{-2j\omega k} f(t+k)f(t-k) \qquad (6)$$

The equation (6) is useful in the development of the Wigner distribution for the image processing

Pseudo Winger distribution:

The inclusion of window in Wigner distribution definition gives Pseudo Wigner distribution in order to reduce the computations. For a real and discrete function $f(t)$ with a window w of duration $2d+1$, Pseudo Winger distribution can be given as

$$PWD_f(t,\omega) = 2\sum_{k=-d}^{d} \cos(2\omega k)w(k)f(t+k)w(-k)f(t-k) \qquad (7)$$

3.2 Generalized Space Frequency Representation for Image Enhancement

To use GSFR for image enhancement, the Wigner distribution function is extended to two-dimensional space. Such an extension results in a four-dimensional distribution function. The function has two space-domain variables x and y, and two frequency domain variables u and v. The extension to 2D space is then

$$WD(x,y,u,v) = \frac{4}{MN} \sum_{l=-N'/2}^{N'/2} \sum_{k=-M'/2}^{M'/2} \cos(\theta)f(x+k,y+l) * f(x-k,y-l) \qquad (8)$$

Where image size is MxN, window size is $M'xN'$, f is the gray-scale function and $\Theta=4\pi[\mu k/M+vl/N]$. In pseudo-Wigner distribution, the main Wigner kernel is multiplied by another kernel like exponential, $e^{-\lambda\|(k,l)\|}$. Then equation (8) becomes

$$WD(x,y,u,v) = \frac{4}{MN} \sum_{l=-N'/2}^{N'/2} \sum_{k=-M'/2}^{M'/2} e^{-\lambda\|(k,l)\|} * \cos(\theta) * f(x+k,y+l) * f(x-k,y-l) \qquad (9)$$

The main objective behind using this kernel is the candidate pixel (x,y) where WD is being calculated should have maximum influence on the calculations, whereas as one goes away from the candidate pixel, its influence should rapidly decay out.

In order to enhance the image range, equation (9) can be scaled by gain factor, α.

Then the equation becomes

$$WD(x,y,u,v) = \frac{4\alpha}{MN} \sum_{l=-N'/2}^{N'/2} \sum_{k=-M'/2}^{M'/2} e^{-\lambda\|(k,l)\|} * \cos(\theta) * f(x+k,y+l) * f(x-k,y-l) \qquad (10)$$

Refer to Vaidya [8] for in depth study of Generalized Space Frequency Representation Distribution for image processing. Equation (10) gives the 1st order Wigner distribution.

3.3 Generalized Space Frequency Representation for Pedestrian Detection

A weighted average edge pixel value of pedestrian is calculated based on pixel's orientation. Pedestrians are vertically oriented objects compared to other object regions in an image. They are associated with vertical edges that are unaffected by perspective projection, as compared to objects like lane markings, buildings, cars and other similar objects. This property is exploited by the system to calculate the pixels in the image, which belong to the boundary of the pedestrian in the image. The weighted average pixel value is determined by using the sobel operators horizontal x and vertical y and calculating the gradient angle of all the edge pixels. The GSFR filter is applied to the image along with the exponential kernel as shown in the equation (10). The values of the pixel obtained, are in the range of 0 to maximum value corresponding to maximum gray level (i.e. 255 in 8-bit gray images). Based on these values obtained the image is divided, into nine bit-planes or layers. The quantization values of these layers are decided based on heuristics, to isolate the pedestrian like object. The first layer has grayscale values ranging from 0 to 30, the second layer has values from 30 to 60, and the third layer has values from 60 to 90 and so on. A connected component labelling is performed on the combination of adjacent layers for object segmentation as the pedestrian like objects fall in one or two adjacent layers. The connected component labelling identifies multiple pedestrian like objects. False detections are eliminated using the aspect ratio (width to height ratio of the segmented region area) and the location. Some obvious false detections are removed using image row and object distance logic. For example: smaller objects detected near to camera i.e. near to the bottom of image and bigger objects farther from camera i.e. top of image are removed knowing the fact that pedestrian near camera can not be smaller and vice versa. If the identified region is located outside the rows corresponding to specific distance range, then the object is classified as random noise. Thus, out of the identified regions, objects other than the pedestrians are discarded.

4   Experimental results

4.1   Experimental Results for Image Enhancement

Following images shows the result of image enhancement using GSFR with varying alpha factor. First image of size 320x240 is a crowded scene captured at evening time and second image of size 712x534 is captured at daytime. For the first image alpha is same for all 3 channels whereas for second image it is different for all channels. The vehicle occupied by fog is clearly visible after applying GSFR.

Fig. 1.1   Input Image
Fig. 1.2   Output Image with Lambda: 3 and Alpha: 368

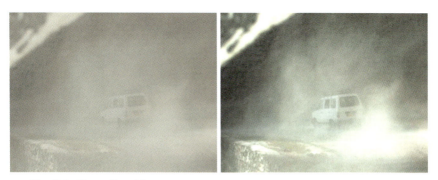

Fig. 2.1   Foggy Input Image

Fig. 2.2   Output Image with Lambda: 0.5 and Alpha: 41 for R channel 21 for
G channel and 48 for B channel

4.2   Experimental Results for Pedestrian Detection

The algorithm is developed in Visual studio and OpenCV [7]. We tested our
algorithm on 4 videos of size 640x480 captured in different scenarios like city
area, dark area with different positions of the pedestrian like pedestrians walking on the road, pedestrians with bag. Figure (3) shows the results of sample
image of size 320x240.

We tested both software modules on a desktop PC of following configuration:
Pentium 4 processor 3 GHz, RAM 1 GB. Table (1) shows the profiling time for
enhancement and pedestrian detection algorithms. Table (2) shows the detection rate for 3 videos.

Fig. 3.1   Input Image

Fig. 3.2   Detected pedestrians

Sr. No.	Software module	Time (ms)
1.	Image Enhancement	124
2.	Pedestrian detection	230

Tab. 1.   Time profiling for image in Figure 2

Video no.	Total No. of frames	Total no of pedestrians	%correct detection	False positives / frame
Clip1	400	575	62.78	0.15
Clip2	300	492	56.50	0.21
Clip3	200	150	73.33	0.13

Tab. 2.   Detection rate

5   Conclusions

The Night vision systems primarily suffer from factors such as low-resolution image, low contrast and high noise due to fog, snow, or rain. We have presented a new technique to provide better visibility. Object or obstacle presented in the scene can be made clear to assist driver. The calculated average detection rate of pedestrian detection algorithm is 63.3% for tested 4 videos. The future work includes tracking the detected pedestrian to reduce profiling time of the pedestrian detection algorithm and trying out different curves like Cardioid, Pear shaped curve for image enhancement.

References

[1]   Yu, Z., et. al., "New image enhancement algorithm for night vision," ICIP 99. Proceedings. 1999 International Conference on Image Processing, Vol. 1,1999.

[2]   Sale, D., et. al., "Superresolution enhancement of night vision image sequences," 2000 IEEE International Conference on Systems, Man, and Cybernetics, Vol. 3, 2000.

[3]   SAITO, H., et. al., "Development of Pedestrian Detection System Using Far-Infrared Ray Camera," SEI Technical Review, Nr. 66, 2008.

[4]   Vinay G. Vaidya, Haralick, R. M., "Wigner Distribution for 2D Motion Estimation Noisy Images," Journal of Visual Communication and Image Representation, Vol. 4, 1993.

[5]   Wigner, E., On The Quantum Correction of Thermodynamic Equilibrium, Phys. Rev 40, 1932.

[6]   Ville, J., Theorie et applications de la notion de signal analytique, Cables et Transmission 2eA 1, 61-74, 1948.

[7]   Source forge, "Open Computer Vision Library" (sourceforge.net/projects/opencvli-brary/)

[8]   Vaidya, V. G., "The Use of Generalized Space Frequency Representation for Motion Estimation from Noisy Image Sequences", Ph.D. Thesis, University of Washington, Seattle, WA, 98045,1992.

Pallavi Kharade, Sanjyot Gindi, Vinay G. Vaidya
KPIT Cummins Infosystems Ltd. Pune
35 and 36, Rajiv Gandhi Infotech Park
Phase – 1, MIDC, Hinjewadi
Pune – 411057, Maharashtra
India
pallavi.kharade@kpitcummins.com
Sanjyot.gindi@kpitcummins.com
vinay.vaidya@kpitcumins.com

Keywords: night vision systems, object detection, wigner distribution, image denoising

Belt-Mic for Phone and In-Vehicle Communication - Pushing Handsfree Audio Performance to the Next Level

K. Rodemer, G. Wessels, Paragon AG

Abstract

Today's handsfree systems for in-vehicle use have reached their performance limits. With microphone arrays, beam forming and sophisticated handsfree audio processing the physical limits of the applied technology are reached. However the quality of the handsfree audio especially as experienced by the far end listener of the communication link is still poor. With the expected introduction of wide band audio in the cellular network, users will experience superior performance when using their cell phone outside the car. This will further raise the expectations on the handsfree audio quality from vehicles. To overcome the physical limits of today's systems a new approach for placing the microphone in the vehicle is required, with the award winning belt-mic system the signal-to-noise ratio can be substantilly improved by putting the microphone close to the speakers mouth. This also opens the door for in-vehicle communication support, i.e. front to rear seat communication.

1   Introduction

Since the introduction of mobile telephony, in vehicles the microphone can be found at a relatively large distance from the voice source (driver or passenger). Preferred mounting locations are the A-pillar, the rear view mirror, the steering wheel column lining, the interior light, or other elements of the cockpit. Sometimes beam forming is used to improve the voice quality, the signals from several microphones are combined by complex beam-forming algorithms to increase the directivity of the microphone system. By removing noise from unwanted directions from the signal also information content in the voice signal is removed. This results in an artificially sounding voice and a poor recognition rate in speech recognition systems. Table 1. shows the signal-to-noise ratio (SNR) for some typical mounting locations in a closed vehicle at 160 km/h with a 50th percentile driver.

	Mounting Location				
	Belt-Mic	Interior light	Rearview mirror	Steering wheel	Dash board
SNR	25 dB	12 dB	13 dB	14 dB	10 dB

Tab. 1.   Signal to noise ratio (SNR) of various mounting locations

Another approach to reduce the distance from the speaker is the use of a goose-neck microphone. However, design considerations and uncomfortable handling prevented the application of gooseneck microphones in passenger cars so far, only in commercial vehicles like busses gooseneck microphones can be found.

To gain an understandable speech signal in convertibles under open roof conditions creates an extreme challenge, especially at velocities above 120 km/h wind turbulences and other driving noise makes microphones at traditional mounting locations in the cockpit useless.

2   Bringing the Microphone to the Speaker

A simple and efficient approach to significantly improve the signal to noise ratio is to move the microphone closer to the sound source - the mouth of the speaker. By reducing the distance from 50 cm to 15 cm an estimated signal gain of 20 lg (15 cm / 50 cm) = 10,45 dB becomes achievable (engery flow per unit area of a spherical surface is proportional to reciprocal of the squared radius).

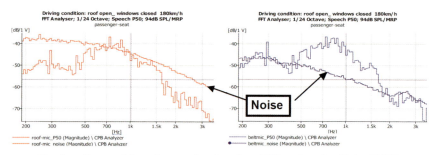

Fig. 1.   Comparision of mounting location, RED = Interior Light, BLUE = belt-mic

Looking for a solution it became soon clear that among the existing components of a vehicle, the safety belt is the only viable approach for moving the microphone to the speaker without using some type of gooseneck microphone.

The safety belt has a stable connection to the vehicle and is mounted in every new car. Now the challenge was to integrate the microphones into the belt, without impacting function, performance or comfort of the belt.

In Fig. 1. a comparison is shown of the driving noise simultaneously recorded at two microphone mounting locations, at a speed of 190 km/h, in a convertible with open roof and an artificial voice (ITU-T P.50), used for objective test methods for speech communication systems (Fig 2. shows the test set-up).

Fig. 2.   Test set-up

2.1   Putting the Microphone in the Belt

It was clear that the car manufacturers would only accept a microphone in the belt if it is proven that safety, usability and comfort of the belt are maintained. For the backside of the belt an even surface must be kept. This lead to the requirement to minimize the size of the microphone unit and to integrate the signal lines into the belt. A safety belt is exposed to relatively harsh wear and use conditions. Nevertheless the function has to be guaranteed over the lifetime of the belt.

To satisfy these requirements a sophisticated mounting technique for the microphone capsules was developed. The signal lines are made of a special alloy and a flexible structure of the wires are woven into the belt. With this approach the signal lines gain flexibility and the belt-mic survives more than 120,000 reversed bend cycles over the seat belt deflection fitting. The microphone unit itself had to be designed to meet the tightness requirements for passing salt water spray tests. Fig 3. shows the microphone capsule.

2.2 First Applications

Handsfree voice quality as experienced by the far end listener is still poor when used in noisy environments especially when the vehicle performance is compared with stationary systems as available today in offices or homes. Even for premium vehicles which are relatively quiet the improvement of 10 dB in SNR leads to a tremendous improvement for the far end listener. As already mentioned it is especially in convertibles and sports cars with a loud engine noise difficult to use hands-free, or speech recognition systems at all with traditional microphone mounting locations.

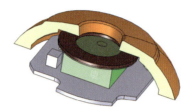

Fig. 3.   Microphone capsule

Main sources of noise include:
- ▶ Wind noise and turbulences under open roof or open window conditions
- ▶ Engine noise
- ▶ Rolling noise
- ▶ Air ventilation noise
- ▶ Pass-by noise from the oncoming traffic
- ▶ Sound echoes from road side constructions

The drivers of convertibles either have to use aids like head sets or have to reduce the speed to be heard and understood by their far end communication partner. The automobile industry has been looking for a solution for this user group since a long time. In light of this situation it is not astonishing that the Audi R8 Spyder and Coupe, a convertible and a sports car, were the first serial production vehicles which were equipped with a belt-mic.

A signal-to-noise improvement of more than 10 dB could be achieved in this car. Since the belt-mic is in the near field of the sound source, the low frequency sensitivity is improved, the information content in the voice signal is increased, which helps to further improve the recognition rate of a speech dialog system.

Depending on the body size of an occupant the microphone position varies with the belt extension. It turned out that by placing 3 microphones on the belt with a distance of appr. 15 cm between them, 95% of the user population are optimally covered, for the remaining 5% the maximum loss in SNR is less than 3 dB. An intelligent selection algorithm was developed to always use the best microphone. Special damping measures in the capsule design and in the signal processing software were taken to eliminate the effects of wind turbulences from the driving air flow or the air ventilation flow.

Finally abuse conditions had to be considered, e.g. a microphone is covered by hairs or clothes, or the microphone experiences scratching with fingers or scrubbing with clothes. In these cases the selection algorithm switches to the next best microphone, with an SNR degradation of less than 3 dB and still significantly better than traditional microphone locations.

3 Processing of the Belt-Mic Signal - the Voice Gateway (VGW)

3.1 Microphone Selection

The basic principle in the current generation is to always select the microphone with the best SNR, using an beam forming approach will be one of the next steps to further improve the signal. A single criteria for microphone selection like the energy content of the signal provides already a basic solution, however windfall, scratching, or other influences could cause unnecessary switching between the microphones. Therefore an algorithm was developed which uses additional criteria to distinguish speech from noise, main criteria are correlation coefficient, signal run time, and energy content. With this approach a reaction time of less than 150 ms for selecting the optimal microphone can be achieved. Fig 4. shows the decision tree at the heart of the algorithm.

3.2 Echo Compensation and Noise Reduction

The special features of the belt-mic system create some new requirements for the signal processing of handsfree systems. A traditional handsfree system provides duplex capability with echo cancellation and noise. For a belt-mic system this architecture has to deviate in some points to address some special characteristics of the belt-mic system, i.e. the possibility that the microphone moves with the breathing of the passenger and the need in convertibles to quickly switch between microphones. Fig 5. shows the first generation solution which provides already a significant performance improvement especially under open roof conditions. Fig 6. shows the development underway for the next product generation, from which hands-free telephony and speech recognition will benefit.

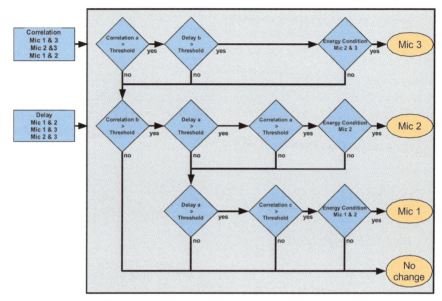

Fig. 4. Decision tree

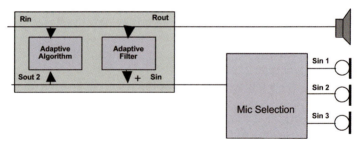

Fig. 5. First generation system

3.2 In-Vehicle Communication

Most drivers have experienced that sometimes they have to turn their heads to be understood by the rear seat occupants, mostly due to the background noise especially at high speeds. Therefore an inter vehicle communication link would not also improve the audibility but also significantly increase the safety. Critical parameters for the performance of these systems are signal-to-noise ratio and signal run time from mouth to microphone. Especially in these two categories microphones in the belt have a significant advantage over other microphone mounting location, e.g. the roof. For convertibles the belt is in fact the only location at which electronically enhanced in-vehicle communication under open roof conditions is possible.

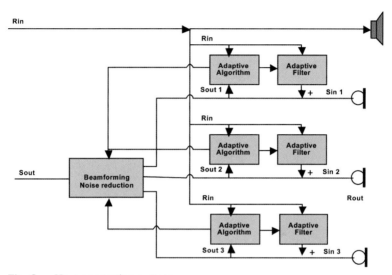

Fig. 6.   Next generation system

4   Connecting the Belt-Mics to the VGW

The current generation belt-mic is connected via 2 analog lines per microphone, in total 6 lines per belt to the audio processing unit. Future high-end systems will use belt-mics on all seats to support in vehicle communication and hands-free phone calls. Direct analog connection of each microphone can be effective for up to 4 seats as long as the processing unit is located close to belt connectors near the tunnel. However the vehicle designers would prefer to freely position the processing unit and to reduce copper wires, therefore paragon currently develops a 2-wire digital audio link between the belts and the processing unit (see Fig 7. for a schematic of the interconnection wiring). The 2 wire connection is used for signalling and power connection. A bus or ring topology can be used, where the ring topology will provide some advantages in terms of fail safe operation.

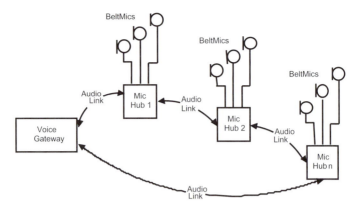

Fig. 7.   Digital Audio Link

5   Putting the System Together

With the special requirements for audio processing and the required CPU load for inter-vehicle communication it makes sense to change the audio processing architecture in the vehicle and move the echo compensation and noise reduction and other supporting functions for hands free telephony and speech recognition into the voice gateway as shown in Fig 8.

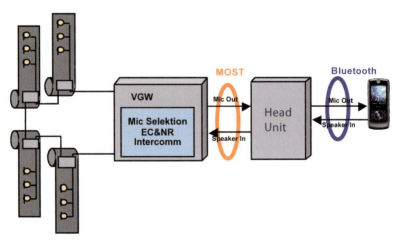

Fig. 8.   System view

6   Conclusions

Above all highly sophisticated audio processing algorithms the signal-to-noise challenge in vehicles can be very simply reduced by putting the microphone closer to the speaker. The belt-mic solves this in a very elegant way: it is easy to add an belt-mic to existing vehicle architectures and it simplifies and eases the interior-design by removing microphones from the roof or rear view mirror bracket. Hands-free speech quality and recognition rate of speech dialog systems are significantly improved. By using a bi-directional digital audio connection technology the system is scalable from 2 to 7 seat vehicles. Comfort and safety under all driving conditions are improved.

Klaus Rodemer
paragon AG
Schwalbenweg 29
33129 Delbrück
Germany
klaus.rodemer@paragon-online.de

Gerhard Wessels
paragon AG
Nordostpark 9
90411 Nürnberg
Germany
gerhard.wessels@paragon-online.de

Keywords: belt-mic, belt, microphone, handsfree, digital audio link, microphone selection, in-vehicle communication

HMM Based Autarkic Reconstruction of Motorcycle Behavior from Low-Cost Inertial Measurements

N. Munzinger, R. Filliger, S. Bays, K. Hug, Berne University of Applied Sciences

Abstract

Based on autarkic data from a low-cost, 6-axes inertial measurement unit (IMU), which is fixed onto and power-supplied by the motorcycles battery, we reconstruct forward velocity and elementary driving behavior of a motorcycle using a Hidden Markov Model (HMM). The notorious drift problem of integrated IMU data is mastered by using the voltage fluctuations of the motorcycle's battery as a stabilizing external signal. Despite the structural simplicity of the algorithm and the relatively low performance of the IMU, the proposed off-line estimator is, after a short learning phase, accurate for a large class of motorcycles.

1 Introduction

The economic and social welfare of any modern society strongly depends on the capacities of the available vehicular traffic systems and the effective driving behavior of its users. Gaining objective insight into the actual driving behavior of traffic network users is therefore a crucial issue in order to enhance road safety and driver efficiency. This issue is partially addressed in the upcoming "Black-Box" mass technology for four wheelers which has already considerable impact on the car insurance industry [1]. For powered two wheelers, no Black-Box technology is available which respects both the error bounds on forward speed imposed by classical accident reconstruction and the financial constraints of mass technologies. The gap prompted our choice of topic and the exposed results constitute a first step towards commercial development of Black-Box technology for motorcycles.

The aircraft's flight data recorder is an example of Black-Box technology which provides objective information regarding time, speed and tracking factors as they relate, for example, to the cause of an accident. An important conceptual difference to aircraft flight data recorder concerns the sensitivity of individual car drivers and motorists with respect to over-controlling their privacy. This is why Black-Box solutions using global positioning systems – as proposed e.g. in

[2] – should be excluded. Indeed, GPS solutions are able to record driving patterns (such as where and when someone drives) which do have privacy rights. In order to use and transmit recorded information without special care, the Black-Box should gather only local data such as speed, braking activities and changes in acceleration.

Accordingly we focus on Black-Box technology for motorists delivering only forward speed, together with a qualitative description of the trajectory, consisting of different cornering- (right turn, straight on, left turn) and tracking (acceleration, constant run, braking) states. These pieces of information are computed from measurements of a commercially available, low-cost 6-axis IMU (3 acceleration axes and the angular rates of the 3 axes) fixed onto the battery of the motorcycle. One of our results is that the battery plays a central role for the low-cost solution. Indeed, besides mechanical support, the battery also provides power supply and – indirectly via the voltage fluctuations delivered from the generator – an external speed signal from the motor which solves the notorious drift problem of integrated IMU data.

As for the analytical approach for state estimation, we will adopt the specifically suited Hidden Markov Model (HMM). The use of HMMs allows building flexible and simple state estimators which can be calibrated and trained on every motorcycle individually. We are not aware of any comparable study for motorcycle driving behavior. Especially the reconstruction of forward speed using the fusion of battery voltage fluctuations and IMU data seems entirely new. The HMM approach is a classical state-estimator tool in the engineering sciences and was successfully applied to many fields including traffic engineering for cars. For example, Liu and Kuge [3] showed how to model road-positioning by a series of HMMs and Kumagaï et al. [4] studied dangerous situations at crossroads, by predicting driver actions using HMMs.

2 Experimental Setup

Fig. 1 shows on the left hand side (LHS) the onboard data acquisition (hardware, attached onto the motorcycle) and, on the right hand side (RHS), the offline data processing (software). In addition, the setups for two reference measurements – GPS for vehicle speed reference and Hall sensors for front- and rear-wheel speed references – are indicated.

2.1 On Board Data Acquisition

The onboard data acquisition (LHS in Fig. 1) contains (i) a 6-axis, micro elec-
tromechanical IMU, measuring specific forces along- and angular rates around
three orthogonal axes, and (ii) a shunt measuring the voltage of the motorcycle's battery. The latter is an oscillating voltage signal whose fundamental
frequency is proportional to the generator's angular rate. An initial calibration
procedure yields the constant of proportionality between the motor's- and the
generator's angular rates and allows computing the RPMs of the motor (here-
after called the motor speed). All data is synchronized and stored on a compact
flash memory card.

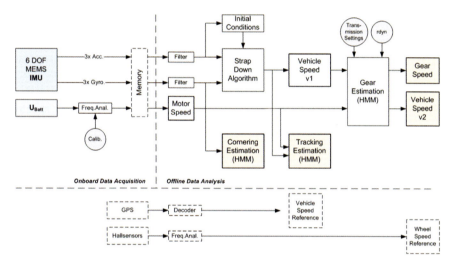

Fig. 1. The experimental setup, described in the main text, consists of an
on board data acquisition device (LHS, see section 2.1), an offline
data analysis (RHS, see section 2.2) and GPS- and Hall-sensors
based reference measurements. The main outputs of the signal
processing algorithms are gear speed, vehicle speed v2, cornering-
and tracking states.

2.2 Offline Data Analysis

The offline data analysis (RHS in Fig. 1) generates the output by using minimum knowledge of motorcycle parameters. We in fact only need the gearbox
settings and the dynamic roll-radius (called rdyn in Fig. 1) of the rear wheel.
The quantitative output provides vehicle and gear speed. The qualitative output gives cornering and tracking states.

The accelerations (more precisely, the specific forces) and angular rates expressed in the body fixed coordinate system as delivered by the IMU are low-pass filtered, re-sampled at 100 Hz and enter a standard navigation algorithm (see e.g., [5] Section 3.3.2). The initial orientation of the IMU, needed for the navigation algorithm, is estimated using measurement data from a motion free period of length of two seconds. The navigation algorithms free the measured specific forces from the influence of gravity (hence only dynamic accelerations remain) and deliver, through numerical integration, objective speed information in the vehicle's longitudinal and lateral direction.

Because of notorious drifts and noise in low-cost IMU acceleration measurements, the estimated vehicle speed is – compared to the GPS speed reference – not accurate enough. Another speed information, the motor speed (in rounds per minute), is used to stabilize the IMU-drift. The optimal fusion of these two independent pieces of speed information uses the motorcycles gearbox settings and yields as the output of a HMM (see section 2.3), the 20 Hz sequence of the most probable gear speed states (e.g. idle, 1, 2, ..., 6) of the motorcycle. From this sequence of gear speed states, the motorcycle's forward speed – probably the most important piece of information for accident reconstruction – is computed using the motor speed and the roll-radius of the rear wheel.

2.3 HMM Based Evaluations

Our ignorance of both the driver's actions and the actual road and traffic conditions lead us to model the actual gear speed and the cornering and tracking states as Markov chains. Recall that in a regular Markov model, it is supposed that the actual state of the process (here, the gear speed, cornering and tracking states) is directly visible to the observer. Therefore the transition matrices are the only parameters necessary to fix the statistics of the future, given the present. However, these states are not directly accessible to us as we only get noisy IMU and voltage measurements. They are therefore related to the true states through conditional probabilities. These conditional probabilities are learned iteratively with the Baum-Welch expectation-maximization algorithm using some training data [6]. Once all the probabilities are defined, the state analyzer then computes the most likely sequence of motorcycle states for a given sequence of observations using standard HMM-algorithms [6].

To be more explicit, we model the cornering state of the motorcycle at some given discrete point in time through one of the following: left cornering, right cornering, straight run; and the tracking state through one of the following: acceleration, deceleration (braking), constant run and gear speed change.

The latter state (gear speed change) effectively indicates moments of undefined states during clutch actions.

To infer the cornering state, the angular rate of the motorcycle's vertical axis is used and to infer the tracking states, the changes in vehicle- and motor speed are used as observation variables. As indicated in Fig. 1, the actual gear speed state is estimated by observing IMU-based vehicle speed v1 and motor speed. We in fact compare the ratio between motor- and vehicle speed v1 with the constant transmission ratios of the motorcycle's drive train (commonly available from the motorcycle's data sheet) and select the corresponding, most probable gear speed state. Proceeding from the sequence of most probable gear speed states we then compute the new vehicle speed information v2, using motor speed and the dynamic roll radius of the rear-wheel.

3   Results

In this section, one generic maneuver on public roadways is measured and reconstructed. Our test-motorcycle is a 1993 Suzuki GSX750 F, propelled by a 748 ccm, 4 cylinder four-stroke engine and weighs about 230 kg (without rider). As a side-remark we note that an analogous analysis with a scooter yielding results of comparable accuracy has been performed. Fig. 2 shows the 180° turn-around maneuver, subdivided into nine pieces (1)-(9), with several time- and speed indications. The speed estimates will be compared to the GPS speed reference.

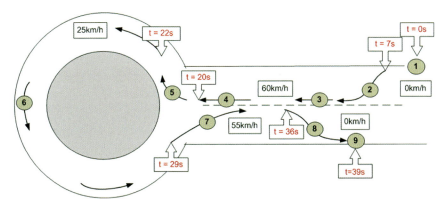

Fig. 2.   180° turn-around manoeuvre in a roundabout, subdivided into nine pieces with several time and speed indications

3.1 Estimate of Vehicle Speed V1 (Without Motor Speed)

The top left drawing in Fig. 3 shows the vehicle speed based on the inertial measurements alone, together with the GPS reference and its 5%-error bounds above 10km/h. After phases of accelerations (2), (3) and decelerations (4), the passage of the roundabout between 20s and 30s at roughly constant speed is visible (5), (6). The acceleration towards the end of phase (6) indicates leaving the roundabout (7), followed by the deceleration (8) which indicates the end of the maneuver. With respect to GPS reference, the vehicle speed estimation v1, which is only IMU-based, stays not within 5% error bounds and therefore, should be improved.

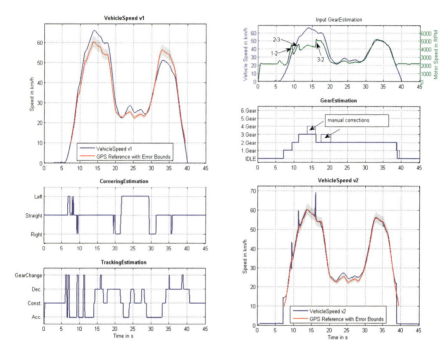

Fig. 3.   HMM-output based on the manoeuvre given in Fig. 2. The description is in the main text.

3.2 Cornering Estimates

The middle left drawing in Fig. 3 shows the result of the HMM cornering estimation. All states are correctly reproduced: left-right maneuver at 7s placing the motorcycle on the road (2), alignment with traffic flow direction (3), (4), as well as entry into (5), turning (6) and exit out of the roundabout (7) between

20 s and 30 s and finally a short right turn, which places the motorcycle on the curbside (8) at 35 s.

3.3 Tracking Estimates

The bottom left drawing in Fig. 3 shows the HMM estimated tracking states. Phases of acceleration, constant drive or deceleration are well reproduced. A fourth state namely gear change, indicates moments of undefined clutch states: at 7 s, the phase of start-up (engaging) is recognized by the model. At 9 s, 11 s, and 16 s, three gear speed changes are visible. The gear speed change at 35s is not obvious, but the disengaging at 35 s is well recognized.

3.4 Gear Speed Estimate with Manual Corrections

The top right drawing in Fig. 3 shows, on different scales, the HMM inputs necessary to estimate the most probable sequence of gear speed states name-ly, vehicle speed and motor speed. The resulting sequence of gear speeds – depicted in the middle right drawing in Fig. 3 – is, apart the sequences between 13 s-15 s and 17 s-20 s, correctly estimated. It is worth noting that erroneous gear speed estimations, as the ones occurring here, are easily detectable. Indeed, the superposition of the vehicle speed and motor speed in the top right drawing gives clear indications about gear speed changes. We easily detect for example two gear speed changes during the acceleration phase (from state 1 to 2 at 9 s and from state 2 to 3 at 11 s), and a third one (back from state 3 to 2 at 16 s) shortly before the deceleration (4). Based on this (non-Markovian) evidence, the HMM-gear speed estimation can be manually corrected if necessary. For the here presented maneuver, two manual correc-tions of the sequence of gear speed states have been applied (the corrections are indicated in the drawing). The resulting bold blue line is now in perfect agreement with the true sequence of gear speed states.

3.5 Estimate of Vehicle Speed V2 (with Motor Speed)

The bottom right drawing in Fig. 3 shows the computed wheel speed, based on the corrected HMM sequence of most probable gear speed states, the motor speed and the roll radius. Above 15 km/h, the wheel speed estimation is in almost perfect agreement with the GPS reference. The three jumps at 9 s, 11 s and 16 s indicate gear speed changes. Below 15km/h, IMU drift and undefined clutch states (motor idle and clutch disengaged or phases of engaging) influ-ence remarkably the probabilistic estimation of the gear speed and may lead to erroneous wheel speed estimations. Moreover, during periods of important cornering (6), the wheel speed touches the 5%-error bounds with respect to the GPS reference. This discrepancy occurred simply because during turning, the road/wheel contact does not take place on the largest circumference of the wheel. This well known fact (see [7], section 4.1.3 for an analytical treatment) enhances the angular velocity of the wheel, without increasing forward speed of the motorcycle.

Another source of differences between the rear wheel speed and the vehicle's (true) forward speed is longitudinal wheel slip. The presence of rear wheel slip will overestimate forward speed during acceleration and will underestimate forward speed during periods of rear wheel braking. For non-critical maneuvers, this error is clearly below 5%.

Note that critical maneuvers (full braking, sharp cornering or combinations) are detected by the qualitative output of the HMM. Therefore the off-line analysis is able to recognize situations where the wheel speed is not a faithful piece of information for inferring forward speed.

4   Conclusion

Despite the fact that the dynamics of single tracked vehicles are considerably more difficult to describe than dynamics of double tracked vehicles, an autarkic gross motion trajectory reconstruction of motorcycles is feasible with low cost devices. Basically, a low-cost 6 axis inertial measurement unit (IMU), a shunt and a clock together with a storage device are enough to compose embedded black-box technology for motorcycles. Power-supply and the IMU-drift stabilizing external voltage signal are both delivered by the motorcycles battery. With a minimum amount of expert knowledge, an initial transmission calibration and a few commercially available motorcycle parameters, we can reconstruct forward speed as well as a cornering- and tracking states with high accuracy. In particular, during phases without important accelerations and cornering, the forward speed estimation almost coincides with GPS-speed and errors stay below 5%. During phases with important accelerations or cornering the estimation errors may grow beyond the 5% error bounds. Such critical situations however, are easily detectable and partially removable through ad-hoc considerations.

References

[1]   Palmer, S., Black-Box technology and its implications to the Auto-Insurance Industry, Communication of the Injury Sciences LLC, 2002.

[2]   Waegli, A. et al., Accurate Trajectory and Orientation of a Motorcycle derived from low-cost Satellite and Inertial Measurement Systems, Proceedings of 7th ISEA CONFERENCE Biarritz, 2008.

[3]   Liu, A., Pentland, A. P., Modelling and prediction of human behavior, Neural Computation, Vol. 11, 229-242, 1999.

[4]   Kumagai, T. et al., Prediction of driving behavior through probabilistic inference. Proceedings of the Eight International Conference on Engineering Applications of Neural Networks, 2003.

[5]   Wendel, J., Integrierte Navigationssysteme, Oldenbourg, München, 2007.

[6]   Rabiner L. R., A. Tutorial on Hidden Markov Models and Selected Applications in Speech Recognition, Proc. of the IEEE, Vol. 77, 257-286, 1989.

[7]   Cossalter, V., Motorcycle Dynamics, LULU press, Italy, 2006.

Nathan Munzinger, Roger Filliger, Simon Bays, Kurt Hug
Berner Fachhochschule TI
Quellgasse 21
2501 Biel
Switzerland
mgn2@bfh.ch
roger.filliger@bfh.ch
bas8@bfh.ch
kurt.hug@bfh.ch

Keywords: HMM, motorcycle dynamics, driving safety, accident reconstruction

Phase Spread Segmentation of Pedestrians in Far Infrared Images

D. Olmeda, A. de la Escalera, J. M. Armingol, Universidad Carlos III de Madrid

Abstract

This article describes a methodology for extracting interesting areas in far infrared (FIR) images that may contain pedestrians. It is part of a larger set of algorithms that are part of an Advanced Driver Assistance System (ADAS). The grey level of an object in a FIR image can shift due to changes of the sensor's temperature. In this paper a contrast and luminance invariant method based on the phase congruency of the signal is proposed. The image is exhaustively searched for regions that may contain a pedestrian based on local phase symmetry at different scales and orientation. Areas with high probability are then feed to a subsequent classification step. By applying this method large areas of the image can be safely ignored, reducing the computation time of the classifier. This method has been tested in the IVVI experimental vehicle in real urban driving scenarios.

1 Introduction

Avoidance of road hazards is an important task of Advanced Driver Assistance Systems (ADAS). In developing this kind of safety systems we must devote special attention to pedestrians, as they are the most fragile element in road environments. However, it is not a trivial task. Pedestrians may have different sizes and shapes, are moving obstacles, and can cross the path of the vehicle unexpectedly at any time. More than 75% of fatal accidents involving a pedestrian happen outside of enabled level crossing [1].

The illumination conditions are a critical factor in road accidents involving a pedestrian. More than half of them happen between dusk and dawn or in low visibility conditions such as under heavy rain. Under these circumstances the driver have less time to react to the presence of a pedestrian. In this article the authors present an algorithm for detecting regions of interest in far infrared (FIR) images that may contain a pedestrian. Since thermal cameras do not require external lighting, this system allows the detection of pedestrians at a distance beyond which the driver can see.

The images from microbolometer-based cameras reproduce the magnitude of heat emission by the scene objects that hit the sensor plane. The main advantage over visible light cameras is that there is no need of any illumination in the scene, so they can be used in total darkness or, as in this case, while driving at night. This kind of cameras usually are sensible in a wide spectrum, and this sensibility is greatly determined by the sensor's own temperature. The variation of this temperature shifts the image histogram in a way that is both non-linear and dependent of the specific sensor being used. Another problem is that the temperature difference between the pedestrian and the background is unknown and vary depending on exterior conditions. The main goal of the system here proposed is to extract image features that are invariant to the sensor's temperature changes and that are present within a wide range of scene temperatures.

Some authors propose the selection of candidates by locally thresholding the image, searching for hotspots, or image regions hotter that their background [2]. In practice, it may be the case that the background is hotter than the pedestrian. It is also possible that the external temperature is much lower, distorting the clothing appearance. In [3] a pre-processing step is proposed that compensates for this clothing-based distortion using vertically biased morphological operations.

In this paper, the authors propose a segmentation based on clustering of contiguous regions of the image that share a similar spread of phase in the frequency spectrum. Areas with strong changes in phase delimit the borders of these regions. A distance transformation of these regions builds an energy map for a watershed algorithm. The regions grow and merge until they reach one of these borders.

The algorithm has been tested, as part of a pedestrian detector system aboard the IVVI experimental vehicle, in sequences recorded in urban environments and with different external temperatures, with satisfactory results. This approach allows the algorithm to quickly discard large areas of the image with less probability of containing a pedestrian. The classification step would then only have to compute a reduced set of regions of interest.

This paper is organized as follows. Section 2 details the segmentation of low probability areas of the image. The selection of regions of interest (ROI) that may contain a pedestrian is explained in section 3. To ensure the presence of pedestrians inside these ROIs, a classification step is needed. In section 4 the application of a classification algorithm is briefly described. Finally, conclusions and future work are presented in section 5.

2   Phase Spread

The intensity derivative magnitude will detect features such as borders but, not being a normalized measure, its value rely on the difference of temperature between the object and the background and also on the scale. Local phase of an image is a better basis for far infrared feature detection and segmentation because it is invariant to intensity. It characterizes intensity features based on the shape of the wave, but without worrying about the amplitude of it. The phase of the image is calculated by extracting a set of frequencies of the signal, as a Fourier series expansion, that represents the information of the features that we want to extract. A set of filters is created, each of which extract the information at a narrow range of frequencies. For each point in the image, a measurement of its phase over a range of scales and orientations is calculated.

2.1   Invariant Features

As said before, the goal of this system is to exploit far infrared image features that are invariant to contrast and to the sensor's temperature. This would allow the use of "low-cost", uncalibrated and non-refrigerated microbolometer sensors with a high detection success rate.

▶ Grey Level vs. Sensor Temperature: Objects in far infrared images are represented with a grey level according to their temperature. This grey level is function of the objects temperature, the emission factor of the body and the sensors temperature. The temperature of all pedestrians can be assumed to be constant within a range, as is the emission factor. However, the sensor's temperature changes over time and has an important impact on the appearance of the images captured. As the temperature rises the histogram is shifted in a non-linear way.

▶ Gradient: Gradient of a FIR image can extract the shape of an object against a uniform background. In practice, the temperature of an object is not always evenly distributed across its surface and the background can contain objects at different temperatures. Another drawback of gradient as an image feature is that it isn't scale invariant. In this case, the shape of a pedestrian far away from the camera would have less importance than one close to it.

▶ Phase Congruency: Features of high phase congruency are those in which a wide range of their Fourier components is in phase. That is, we are looking for points in the wave that have an order, but without worrying about the shape or amplitude of it. Phase congruency can extract rich information from images, such as borders, but it is also a normalized magnitude that is contrast invariant. This way, no matter what the contrast between object and background is, equally symmetric edges will have equal weight.

2.2 Phase Spread Features

Regions of high phase congruency correspond to points in the wave with a high slope or at peaks. Decomposition of smooth areas has its frequencies spread over a wider range, thus being its phase congruency score lower.

We need to extract a set of frequencies of the signal that represents the information of the features that we want to extract. The solution is to convolve the Fourier transformed signal with a set of filters. Fig. 1(a) represents the real imaginary parts of the FFT. Each of these filter extract the information at a narrow range of frequencies. Because the filters have to be used over a complex signal, they have to be complex too. In this case, we use a set of Gabor filters. The even part of the filter is a sine curve and the odd part a cosine. Both signals are convoluted with a Gaussian of the same variance. The one-dimensional Gabor filter can be extended into two dimensions by applying another Gaussian function across the filter perpendicular to its orientation.

The phase congruency implementation used is derived from the work of Kovesi [4]. Fig. 1(b) contains examples of these filters at different scales. Applying each of these filters results in an image with only a reduced range of frequencies.

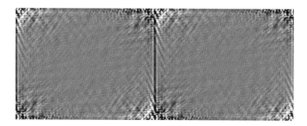

Fig. 1.   (a, top) Fourier transform of the far infrared image, (b, bottom) band-pass filters

Fig. 2 contains the inverse Fourier transform of the filtered images. Each frequency range is known as a scale, and for each scale the filter is rotated a number of times. This way, information for multiple orientations can be extracted. In the case of pedestrian richer information can be found in vertical edges.

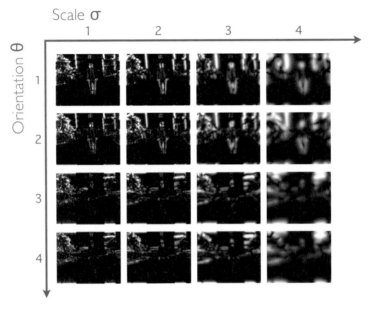

Fig. 2.   Image filtered at different scales and orientations.

The phase congruency score for every pixel in the image is determined by equation 1, where A_n is the square mean of the convolution of the signal with the odd and even filter; n is the order of the Fourier expansion, w a constant, ϕ_n is the phase offset and 0 is the weighted mean of all Fourier terms that maximizes the equation. The magnitude of maximums of phase congruency score of a test image can be seen in Fig. 3b.

$$PC(x) = \max_{\theta \in [0,2\pi]} \frac{\sum_n A_n \cos(n\omega x + \phi_n - \theta)}{\sum_n A_n} \tag{1}$$

Regions of the image without significant phase transitions have a low phase congruency score. On the other hand, regions with high scores are spatially distinguishable from their neighbours. Smaller regions with constant phase congruency values are more likely to contain useful information of borders belonging to pedestrians. The relevance of each pixel is set to be the inverse of its distance to the closer border. To evaluate the weight of each pixel a watershed segmentation [5] is performed over the distance transform of the magnitude image of the phase congruency. As an example, Fig. 3d contains a watershed segmentation of the distance transform Fig. 3c.

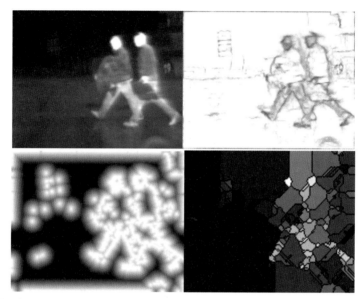

Fig. 3. (a, top left) Original Far Infrared image, (b, top right) Maximums of phase congruency, (c, bottom left) Distance transform, (d, bottom right) Watershed Clustering

3 Selection of Regions of Interest

Pedestrians in images can appear at different scales, depending on their distance to the camera. However, an exhaustive search and classification of every possible region of interest not appropriate. In this section a fast way to discard areas with low probability of containing a pedestrian is detailed.

First, a set of regions of interest is created as rectangular boxes with a aspect ratio of 1/2. To avoid searching for pedestrian in unlikely areas some geometric restrictions are applied. Only pedestrians on the ground plane and inside the vehicle's trajectory are looked for. Knowing the intrinsic parameters of the pinhole modelled camera and its position and orientation over the ground plane it is possible to establish an homography projection of the ground plane over the sensor plane. The regions of interest are created at fixed ranges of distances to the camera. The world system of coordinates is placed on the ground plane, moving along with the vehicle and so does the camera position. The changes on the extrinsic parameters due to vibrations are captured by means of an inertial navigation system with three gyroscopes and three accelerometers.

As explained before, the pixels of the watershed image have a weight based on the size of the blob they belong to. Bigger blobs have lower weights, as they are smooth areas with less important information. The score of each region of interest is the sum of the weight of every pixel in it, normalized by the size of the ROI (equation 2).

$$S = \frac{\sum_{x=0}^{w} \sum_{y=0}^{h} \phi_{x,y}}{w \cdot h} \tag{2}$$

Where S is the score of the ROI, w is the width, h is the height and ϕ is the weight of each pixel. If the score of the box is below a certain threshold, that region can be ignored and won´t be fed to the classifier.

In Fig. 4 only the surviving ROIs with a high score are represented. Only this reduced subset will be further processed, thus reducing the processing time.

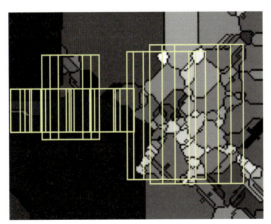

Fig. 4.   Regions of interest that may contain a pedestrian

4   Pedestrian Classification

Final verification of the extracted regions is done by means of grey scale corre-lation with some precomputed models. In far infrared images the most recognizable feature of pedestrians is the silhouette of the body temperature against the background. So, the correlation takes place between the ROIs thresholded with the lowest temperature set for the human body and the models. From several processed sequences, extracted ROIs containing potential pedestrians are manually classified. The models are created computing the mean of the value of each pixel for the training group.

Pedestrians have very different appearances depending on the their gait cycle. The main difference is due to the position of the legs. That's why ROIs containing pedestrians are grouped into four different categories, attending to that position. This approach enables the algorithm to correctly identify a wider diversity of shapes but it takes longer to process four correlations for each candidate. To reduce the number of calculations a fifth model is created for a common characteristic of pedestrians: the head.

To capture the differences between pedestrians located both near and far from the camera, the training set is also split into subgroups of similar sizes. During the detection phase of the algorithm, correlation will only take place between the candidate and the set of models created for the particular range of distances in which the candidate is. An extended description of the classification algorithm can be found in [6].

5   Conclusions and Future Work

In this paper a computer vision algorithm for locating regions of interest in far infrared images with high probability of containing a pedestrian has been presented. The algorithm groups and discards areas of the images without borders. For the extraction of these edges the authors propose the use of phase congruency features over gradient-based ones, as it is invariant to illumination and contrast of the image. This allows the algorithm to also be invariant to the temperature of the camera sensor.

The threshold for the ROI score is set so that false positives ratio is always lower than 10^{-4} FPPW. Under this restriction the algorithm is able to correctly reject 75% of bounding boxes, so that only the remaining 25% of them pass to the classification step.

The experiments were conducted on the IVVI (Intelligent Vehicle Based on Visual Information) in sequences recorded in urban environments and preferably at night, though the method has proven robust for more demanding temperature conditions. The camera used is an Indigo Omega with sensitivity between 7.5μm and 13.5μm. It has a pixel resolution of 164x129 and a grey level depth of 14 bits.

Fig. 5.   The IVVI experimental vehicle

Acknowledgements

This work was supported by the Spanish Government through the Cicyt projects FEDORA (GRANT TRA2010-20225-C03-01), VIDAS-Driver (GRANT TRA2010-21371-C03-02).

References

[1]   National Highway Traffic Safety Administration. Pedestrian statistics. Website, 2009. http://www-fars.nhtsa.dot.gov/ People/PeoplePedestrians.aspx

[2]   Xu, F., Liu, X., Fujimura, K., "Pedestrian Detection and Tracking With Night Vision", IEEE Transactions on Intelligent Transportation Systems, 2005.

[3]   O'Malley, R., Jones, E., Glavin, M. "Detection of pedestrians in far-infrared automotive night vision using region-growing and clothing distortion compensation", Infrared Physics and Technology, Volume 53, 439-449, 2010.

[4]   Kovesi, P., "Image features from phase congruency, Videre: Journal of Computer Vision Research, Volume 1, Pages 1-26, 1999.

[5]   Meyer, F., "Topographic distance and watershed lines", Signal Processing , Vol. 38, pp. 113-125, 1994.

[6]   Olmeda, D., de la Escalera, A., Armingol, J.M., "Far infrared pedestrian detection and tracking for night driving", Robotica, 2010.

Daniel Olmeda, Arturo de la Escalera, José Maria Armingol
Universidad Carlos III de Madrid
C/ Butarque, 15. 28911
Leganés, Madrid
Spain
dolmeda@ing.uc3m.es
escalera@ing.uc3m.es
armingol@ing.uc3m.es

Keywords: far infrared, computer vision, pedestrian detection, advanced driver assistance

Real-Time Pedestrian Recognition in Urban Environments

B. Musleh, A. de la Escalera, J.M. Armingol, University Carlos III of Madrid

Abstract

Traditionally, pedestrian recognition is a great research topic in computer vision applied to advanced driver assistance systems (ADAS); a real-time pedestrian recognition system based on stereo vision is presented in this paper. The most interesting features of the system are that it does not need any extrinsic calibration and it is possible to determine the pedestrian's localization with a bigger resolution than only by using the disparity values. This is possible because the road profile in front of the vehicle is calculated from the v-disparity at each frame. Once the road profile has been generated the obstacles can be classified into elevated obstacles or obstacles on the ground. Regarding the pedestrian recognition, a fast method has been developed based on the similarity between the vertical projection of the pedestrian's silhouette and a normal distribution. Stereo algorithms have a high computation time and in order to cope with this, our algorithm has been implemented in graphics processing unit by means of CUDA.

1 Introduction

In order to have information of depth in computer vision it is necessary to set up two cameras with some ideal characteristics which allow to use the well known equations (1) and (2). These equations can be used to obtain the depth (W) for a point $P = (U, V, W)$ in world coordinates, where its projection over the image plane are (u_L, v_L) for the left camera and (u_R, v_R) for the right camera.

$$W = f \cdot B / \left(u_L - u_R\right) = f \cdot B / d \tag{1}$$

$$v_L / f = V / W; v_R / f = V / W \Rightarrow v_L = v_R \tag{2}$$

Where d is the disparity, f is the focal length and B is the baseline between both cameras. The ideal features are difficult in practice, for this reason a stereo system, the Bumblebee by Pointgrey, is used to rectify the two images.

From the rectified images it is possible to construct the disparity map [1] which represents the depth (W) for every pixel of the image. Once the disparity map has been generated the u-v disparity [2] can be obtained (Fig.1). The v-disparity expresses the histogram over the disparity values for every image row (v coordinate), while the u-disparity does the same but for every column (u coordinate). It is important to note that the obstacle ahead the vehicle and situated perpendicularly, appears as horizontal lines in the u-disparity and as vertical lines in the v-disparity [3] in their corresponding values of disparity. Another interesting feature is that the road profile in front of the vehicle appears as an oblique line in the v-disparity, this feature is very useful because the pitch and the height between the stereo rig and the road can be measured for each frame [4]. For this reason, the system does not need any extrinsic calibration.

Fig. 1.   Visible image and the corresponding disparity map and u-v disparity

Stereo algorithms require a large computing power to be performed in real time, that is why the NVIDIA CUDA [5] is used to process the algorithm on GPU (Graphics Processing Unit). The philosophy of the processing on the GPU lies in using its high degree of parallelization to avoid large loops in the CPU for repetitive tasks.

Section 2 of this paper describes the method of obstacles detection, and their classification is explained in the next section. An improved method of obstacles localization is presented in section 4. In order to evaluate the algorithm, several tests have been performed in urban environments, an example of these tests is commented in the section 5.

2   Obstacles Detection

This stage of the algorithm detects every obstacles in front of the vehicle. The main result is the obstacles map (Fig. 2 (b)), which is composed of every pixel in the disparity map corresponding to obstacles. From the obstacles map, it is possible to determine the regions of interest (ROI) used in a subsequent classification of obstacles (Fig 2 (a)). The ROIs only will be established for obstacles which are inside the study region, which is an area in front of the vehicle to a certain depth in terms of disparity where it is plausible to model the road with a flat geometry. The obstacles detection system follows these tree steps:

▶ The first one is a preliminary detection on the u-disparity based on thresholding, so that the obstacles higher than a threshold measured in pixels are detected. In this way a threshold u-disparity is obtained.

▶ For each thresholded pixel (u , d) of the u-disparity, we study the pixels of column u of the disparity map where, if the value of the disparity is equal to d, then the value of disparity is conserved, but if the value of disparity is different from d, then the value is set at zero. In this way, the obstacles map is generated, that is a dense disparity map where only the pixels of obstacles have values different from zero.

▶ Once the study region has been determined by the minimum disparity to search for obstacles, the obstacles map is thresholded by using this minimum disparity as a threshold to ignore the obstacle outside the study region. As a result of thresholding a binary image is obtained, where a blobs analysis is performed to know different features of the obstacles, such as area and position. Obstacles close to each other are often interpreted by the blobs analysis as a unique obstacle. In order to reduce this problem, the edges of the obstacles are obtained from the obstacles map through a filter and removed from the binary image before the blobs analysis. It is not advisable to remove the edges in the closest area to the vehicle, in order to avoid smashing the obstacles to pieces. It is important to highlight that obstacles with small areas are ignored for the purpose of reducing the computing time.

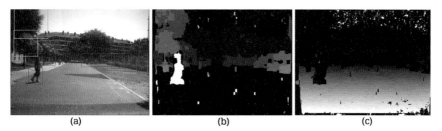

(a)   (b)   (c)

Fig. 2.   (a) Visible image with the regions of interest, (b) the corresponding obstacles map (c) the corresponding free map.

3   Obstacles Classification

In this paper two different classifications of obstacles are presented. The first one distinguishes between obstacles elevated (green rectangles in Fig. 2. (a)) or obstacles on the ground (red rectangles in Fig. 2. (a)) and the second one classifies the obstacles into pedestrians or not.

3.1   Classification of elevated obstacles

In order to classify the obstacles into elevated or non-elevated, it is necessary to know the road profile in front of the vehicle and, as mentioned before, it may be obtained from the v disparity by means of the Hough transform. The road profile relates the v coordinate of the image with the disparity as shown in equation (3).

$$v = m \cdot d + b \qquad\qquad (3)$$

There are situations in urban environments where it is difficult to determine the road profile, usually there are problems when big obstacles or several of them appear ahead of the vehicle [6]. These obstacles can be other vehicles, buildings, walls or when the vehicle come closer to a tunnel. It is possible to get better results if the pixels of the obstacles are eliminated from the disparity map before generating the v-disparity, as a result, it is obtained the free map (Fig. 2. (c)). Fig. 3. shows two results of the computation of the road profile (red line) from the v-disparity in urban environments, in both cases the road profile has been obtained from two different v-disparity. In the first one (left), the obstacles have not been removed and produce an erroneous road profile, and in the second one (right) the road profile is correct because the obstacles are previously removed from the v-disparity.

Once that the road profile has been obtained, it is possible to distinguish between elevated and non-elevated obstacles. Firstly, the theoretical disparity is calculated, and it represents the disparity that an obstacle should have if it was on the ground. For it, the v coordinate at the bottom of the region of interest is used to estimate the theoretical disparity by means of the road profile. In the event of the theoretical disparity was equal to the disparity of the obstacle, then the obstacle is on the ground, but if the theoretical disparity was minor than the disparity of obstacle, then it is elevated.

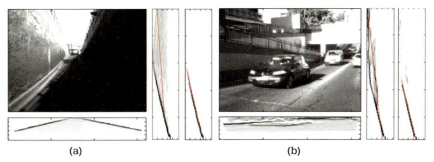

(a)   (b)

Fig. 3.   Two examples of obtaining the road profile from the v-disparity. Visible image, the corresponding disparity map and the u-v disparity

3.2 Pedestrian Classification

This classification divides the obstacles into two groups: pedestrians and non-pedestrians. The results of the classification algorithm is a confidence score for the fact that the obstacle is a pedestrian; it is compared with a threshold and if it is greater, the obstacle is classified as a pedestrian. This classification is based on the similarity between the vertical projection of the silhouette and the histogram of the normal distribution. The vertical projection for each obstacle on the ground is computed by means of the ROIs in the thresholded obstacles map, as a results of the detection stage.

In order to characterize the vertical projection, the standard deviation, σ is computed, as if the vertical projection was the histogram of a normal distribution. In order not to make the standard deviation be a function of the obstacle dimension or independent on the obstacle localization, the standard deviation is divided by the width of the ROI getting σ_w. This standard deviation will be used to compute the score. Several vertical projections of pedestrian have been processed to obtain their standard deviations; these standard deviations follow a normal distribution $N(\mu_{\sigma_w}, \sigma_{\sigma_w})$. In order to compute the score for an obstacle, its standard deviation is used to obtain the value of the probability density function, where the maximum score 100% is produced if the standard deviation is equal to μ_{σ_w} and gets worse when the standard deviation is different from μ_{σ_w}. A scheme of the process is shown in Figure 4.

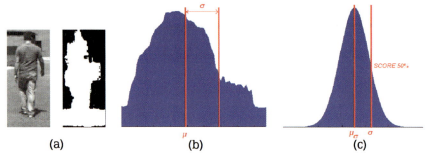

Fig. 4.   Process scheme to obtain a pedestrian score. (a) Pedestrian image
and the corresponding silhouette. (b) The vertical projection of
the pedestrian silhouette. (c) Normal distribution of the standard
deviations and the score for the σ corresponding with the vertical
projection of the pedestrian silhouette.

4   Obstacles Localization

In the case of the obstacles are on the ground, it is possible to determine their
localization with regard to the vehicle with a higher resolution by using the
road profile than if it is elevated because in this last case, only the disparity
information can be used to estimate its localization. The localization of the
obstacles in front of the vehicle in world coordinates (U, W), can be obtained in
terms of the coordinates of the image (u, v). In order to achieve this objective, it
is necessary to combine the stereo equations (4) and the road profile (3), obtaining (5), where C_u corresponds to u coordinate of the optical center.

$$W = f \cdot B / d \qquad U = W \cdot \left(Cu - u \right) / f \qquad (4)$$

$$W = m \cdot f \cdot B / \left(v - b \right) \quad U = m \cdot B \cdot \left(Cu - u \right) / \left(v - b \right) \qquad (5)$$

The Fig. 5. shows a comparison of the resolution of depth (W) between the
two methods of localization. The red line as a result of using only the disparity
values, and the blue line by using the road profile. As the graphic shows, the
resolution using the road profile improves a lot with respect to use the disparity values exclusively. It is important to note that the resolution of the obstacles
is not lineal, that is, the resolution gets worse when the obstacles move away
the vehicle.

5 Experimental Results

Several tests have been performed in urban environments in order to evaluate the robustness and reliability of the proposed detection and classification system. Here we are reporting on a specific test which can be divided into two stages: firstly, a single pedestrian following a zig-zag trajectory and afterwards two pedestrians crossing ahead the vehicle. The Figures 6. (a) and 6. (c) show the results of the obstacles classification for four images in each stage; the pedestrians appear in blue, the elevated obstacles in green and finally, the obstacles on the ground which are not pedestrians appear in red. The Figures 6. (b) and 6. (d) present the results of the pedestrians' localization in both stages, the blue circles represent the localizations obtained by means of the equation (5) and the red crosses represent the localizations obtained by means of the equation (4). The use of the road profile (equation (5)) produce a substantial improvement in the resolution of the obstacles localization, being more significant in areas far from the vehicle.

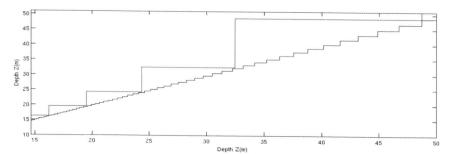

Fig. 5.   Comparison of the resolution between using the disparity values (red line) and and using the road profile (blue line)

Regarding the pedestrian classification, the system has correctly classified the pedestrians in a 78% of the cases with a 4% of false positives. It is important to highlight that the pedestrians have been detected as an obstacle in 97% of the cases, so with this information and if a tracking of the results of pedestrian classification is implemented, it is possible to increase significantly the hit rate of the results of the pedestrians' classification.

The frame rate of the system is upper to 10 images per second because a large part of the algorithm is processed in the GPU through the NVIDIA CUDA, which allows to reduce significantly the computing time with respect to the processing on CPU. For example, the construction of the disparity map is a costly task, the same implementation to generate the disparity map is 15 times faster by using the GPU than with the CPU. It is also important to note that it is possible to use the CPU for other tasks while the GPU is processing. For this comparison, a Core2Duo 2 Ghz and 1 GB of RAM has been used as CPU and a NVIDIA Quadro FX 380 LP as GPU.

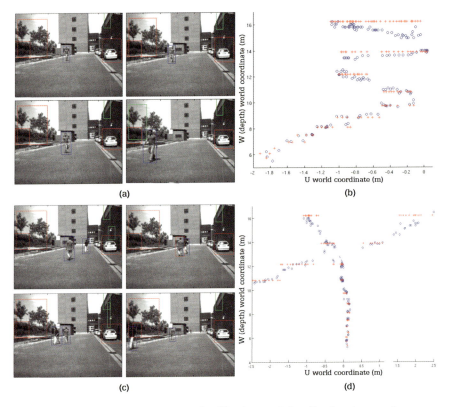

Fig. 6.   Test of pedestrians' classification and localization in a urban environment. (a) and (c) Four images of each stage (zig-zag and crossing) where the ROIs appear over the obstacles. (b) and (d) Pedestrians' localizations as a consequence of the trajectories followed by the pedestrians

6   Acknowledgements

This work also was supported by Spanish government through the CICYT projects FEDORA (Grant TRA2010-20225-C03-01) and VIDAS-DRIVER (Grant TRA2010-21371-C03-02).

References

[1] Scharstein, D, Szeliski, R, A taxonomy and evaluation of dense two-frame stereo correspondence algorithms, International Journal of Computer Vision, 47, 7-42, 2002.

[2] Soquet, N, Perrollaz, M, Labayrade, R, Auber, D, Free space estimation for autonomous navigation, Proceedings of the 5th Int. Conference on Comput. Vision Systems, 1-6, 2007.

[3] Broggi, A, Caraffi, C, Fedriga, R, I, Grisleri, Obstacle detection with stereo vision for off-road vehicle vehicle navigation, Proceedings of the IEEE Conference on Computer vision and Pattern Recognition, 1-6, 2005.

[4] Labayrade, R, Aubert, D, Tarel, J, P, Real time obstacles detection in stereovision on non flat road geometry through V-disparity representation vehicle navigation, Intelligent Vehicle Symposium. 1-6, 2002.

[5] NVIDIA CUDA, Programming guide, 2.3.1 version, NVIDIA Co.

[6] Lee, C,H, Lim Y,C, Kong, S, Lee, J,H, Obstacle localization with a binarized v-disparity map using local maximum frequency values in stereo vision. Proceedings of the International Conference on Signals, Circuits and Systems, 1-4, 2008.

Basam Musleh, Arturo de la Escalera, José María Armingol
University Carlos III of Madrid
Escuela Politécnica Superior C/ Butarque 15
Leganés
Spain
bmusleh@ing.uc3m.es
escalera@ing.uc3m.es
armingol@ing.uc3m.es

Keywords: pedestrian recognition, obstacles detection, U-V disparity, stereo vision

Development of a Low-Cost Automotive Laser Scanner – The EC Project MiniFaros

K. Fürstenberg, F. Ahlers, SICK AG

Abstract

The project "Low cost miniature laser scanner for environmental perception", MiniFaros, is a sensor development project aimed at significantly increasing the penetration of advanced driver assistance systems, ADAS, in the automotive market. A typical top-down approach derives from accidentology to relevant scenarios to selected scenarios, to applications and finally to the laser scanner specifications. In order to demonstrate the MiniFaros' capabilities, 6 safety applications are specified in order to address the most relevant accident scenarios for trucks and passenger cars. They cover three pedestrian related as well as three frontal crash scenarios.

1 Introduction

The MiniFaros consortium stresses that a boost of the market penetration of driver support systems can be realised by generic sensors that are affordable, durable and of compact size to be used in different locations in vehicles or in the infrastructure. Furthermore, these systems need to be based on fully reliable sensor data. All these requirements have not yet met by present day sensors.

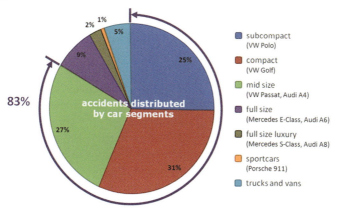

Fig. 1.   Accident distribution by car segment

The wish and determination for considerably fewer accidents have been pronounced by a number of European stakeholders. European Union launched the eSafety Programme in 2002. eSafety is a joint initiative of the European Commission, industry and other stakeholders. This safety initiative aims at accelerating the development and deployment of Intelligent Vehicle Safety Systems (IVSS) that exploit information and communication technologies in intelligent solutions in order to improve road safety. Moreover, EU has launched a programme 'Halving the number of road accident victims in the European Union by 2010: A shared responsibility'.

It is obvious now that these ambitious safety goals will not be met as scheduled. A number of reasons for that can be brought up. One of them is the slow market introduction of IVSS and the high cost of safety applications. Today IVSS are limited to a small part of the premium car segment. Future safety systems must be made affordable to penetrate all vehicle segments since small and medium size cars are dominating the road traffic and thus most of the accidents. The consortium stresses that the improvements have still to go down to basics, i.e. create exploitable sensing systems to boost the enhancement of road safety.

MiniFaros aims at opening up the Advanced Driver Assistance System market for small and medium size cars and broadening the range of possible applications by developing a new low-cost, low power, miniature Laser scanner characterized by high performance. The project shall develop and demonstrate a totally new type of a laser sensor for enhanced environment perception in terms of optics, a solution for replacing the scanning mirror, electronics and the ability to serve numerous automotive applications and even beyond.

2   Technical Objectives

The general objective of MiniFaros is to develop and demonstrate a prototype of a low-cost miniature automotive laser scanner for environment perception. In order to meet this goal, the following technical objectives have been defined for the laser scanner: a) the manufacturing costs shall be low (about 40 €, in mass production), b) it shall be small and compact (about 4 cm x 4 cm x 4 cm, in mass production), c) a MEMS mirror shall be used instead of a macro mechanical scanning system, d) an integrated receiver and time-to-digital-converter shall be used to enable highly precise distance measurements and multi echo technology for optimized bad weather performance, e) highly integrated optical and mechanical components shall be designed to support future low cost mass production, f) improved object detection, tracking and classification algorithms shall be developed and, g) the novel miniature laser scanner shall serve various in-vehicle applications which will be demonstrated at the end of the project.

A main innovation in the MiniFaros project is the combination of omnidirectional lenses and MEMS mirror technology supported by a new electronics solution. Generally the deflection mechanism in laser scanners provides a horizontal angular scanning of the laser spot e.g. rotation of a mirror, to achieve its field of view.

With the novel concept the scanning is not achieved by a horizontal angle deflection, since the deflection mechanism scans the projection of the omnidirectional lens not the real environment. The omnidirectional lens captures the complete field of view at any time. Thus, the deflection mechanism requires both, horizontal and vertical deflection, which will be realised with a MEMS mirror.

3 Requirements

Poor human perception and assessment of traffic situations stands for the largest amount of traffic accident with fatal or severe injury outcome. Several safety functions are developed in order to prevent or mitigate many of these accidents. The system cost for these functions are often relatively high and consequently, the penetration rate of these systems is still low, especially when it comes to smaller cars or commercial vehicles.

In a clear top down approach, MiniFaros started by deriving the selected scenarios from the relevant scenarios for the MiniFaros laser scanner based on accidentology. The selected scenarios will be covered by a set of applications closing with the resulting functional requirements such as detection distance and field of view for the MiniFaros laser scanner.

3.1 Relevant Scenarios

The deliverable D3.1 [1] describes relevant scenarios for the MiniFaros laser scanner based on accident analysis.

Frontal crash accidents are the largest accident type when looking at seriously injured or killed people in Europe. It stands for almost one third of the accidents, 32% for cars and 37% for heavy trucks. It is therefore of high relevance to prevent or mitigate those types.

A second major scenario is involving vulnerable road users (VRUs) including pedestrians. Scenarios where VRUs are involved stand for approximately one fifth of the serious injuries, 18% of the accidents for cars and 24% for heavy trucks.

Accident type (CV: case vehicle)	Case description	Pictogram examples [GIDAS]	Ratio of	
			all frontal accidents [%]	severe or fatal frontal accidents [%]
I. CV hits frontally to another vehicle	Initial speed before the crash is a sum of both vehicles' initial speeds.		22.6	68.4
II. CV driving straight, another vehicle changing to same lane or is already moving in the same lane	Initial speed before the crash is difference of both vehicles' speeds		33.0	23.7
III. CV driving straight, another vehicle is moving with minimal longitudinal speed (e.g. turning, driving out of parking place etc.)	Initial speed is practically speed of the CV		39.2	5.3
IV. CV hits into solid obstacle			5.1	2.6
			99.9	100.0

Tab. 1.   Distribution of frontal crashes (without intersection accidents)

Another relevant scenario is accidents occurring in intersections. Approximately 20% of car accidents and around 18% of truck accident is related to these scenarios. Some of these accidents are on rural roads with higher speeds and out of range for the proposed laser scanner. But urban intersection scenarios are relevant for the MiniFaros laser scanner.

Most of the major scenarios are addressed by the MiniFaros laser scanner and also some of the minor scenarios. In total approximately 54% to 74% of the serious accidents are addressed for cars and 64% to 82% for trucks.

3.2 Selected Scenarios

In order to demonstrate the capability and performance of the novel MiniFaros laser scanner, a limited number of scenarios have been selected. The intention is to address the most relevant scenarios for traffic safety that could be covered by using laser scanner technology. The applications for intersection safety dealing with accidents between vehicles are still under development in other European research projects. Thus they are not adapted in MiniFaros but addressed by the MiniFaros laser scanner.

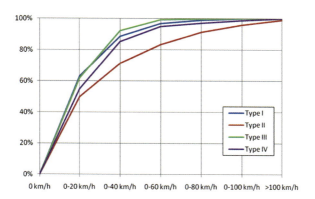

Fig. 2.   Initial speed of CV with regard to frontal crashes

3.3 Functional Requirements

Majority of all frontal crash accidents happen with at least one vehicle hitting frontally. It's only about to judge which vehicle is actually the CV at the moment. The selected scenarios describe the most frequent types of accidents. Accidents happening at junctions when one vehicle is turning and crashes into second vehicle driving straight are not considered.

Initial speeds up to 40 km/h address about 90% of type I,II and IV, which represent 76% of all severe or fatal frontal crash accidents, as illustrated in Figure 2. Thus a detection range for the warning of less than 80 m would be sufficient for the laser scanner.

However, a warning application might not be the suitable solution. Since the warning to the driver needs to be applied more than 3 s before the potential crash many collisions will be avoided by a standard manoeuvre, which was planned by the drivers anyway. Thus a huge number of false warnings will be applied, which will not be accepted by drivers.

Therefore pre-crash systems acting just if the collision is unavoidable are very much in favour. The pre-crash system will mitigate the consequences of accidents, if e.g. an automatic emergency braking system is applied. Table 2 lists the resulting range of the laser scanner in dependence of host-vehicle's and POV's initial speeds for pre-crash applications. The required range to address more than 90% (40 km/h) of the frontal accidents is 30 m.

Required detection distance							
	opponent vehicle						
velocity [kph]	10	20	30	40	50	60	70
10	4	6	8	10	13	15	17
20	7	10	13	16	19	22	25
30	11	15	19	22	26	30	34
40	16	21	25	30	34	39	43
50	22	27	32	38	43	48	53
60	28	34	40	46	52	58	64
70	35	42	49	56	63	69	76

host vehicle

Tab. 2.   Required range of the laser scanner in dependence of the initial speed for the pre-crash application

3.4   Selected Applications

The pedestrian protection application is designed to avoid collisions with pedestrians by warning the driver or mitigate consequences for pedestrians being hit by passenger cars or trucks.

The shapes of the region of warning (ROW) and region of no escape (RONE) both are depending on the host-vehicle speed and yaw rate, as well as on the moving direction and the speed of the endangered pedestrians.

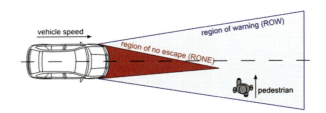

Fig. 3.   Pedestrian protection application for cars

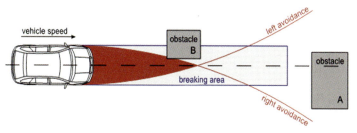

Fig. 4.   Pre-crash application for cars

The pre-crash application (PCA) is designed to predict unavoidable crashes with solid objects and reduce the consequences of the impact. The objects could either be other vehicles, such as passenger cars, trucks or motor cycles or static objects such as traffic light poles, crash barriers or even walls. The measures could be to prefill the brake system, firing the belt pretentioner as well as pre-firing the airbag.The safe distance application (SDA) is an application, providing the driver with information about the clearance in time to the preceding vehicle. If the clearance is falling below a defined threshold, a warning is initiated.

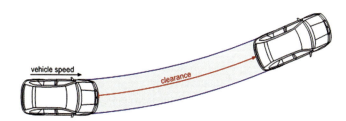

Fig. 5:   Safe distance application for cars

The stop and go application (S&G) will handle acceleration and braking to keep the vehicle at a safe distance to other vehicles in front. This will be possible in low speeds and in dense traffic down to a full stop of the vehicle.

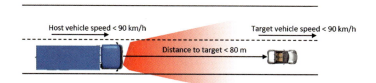

Fig. 6.   Stop and go application for trucks

The start inhibit application (SIA) will prevent the driver from taking off from stationary when there is road users or other object detected close in front of the vehicle. The system actually prevents the vehicle from accelerating.

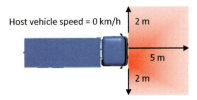

Fig. 7.   Start inhibit application for trucks

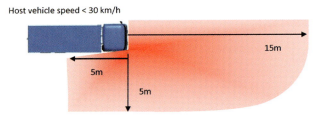

Fig. 8.   Right turn assistance for trucks

The right turn assistance application (RTA) focus on accidents with vulnerable road users such as cyclists and pedestrian in a right turn scenario. Pedestrians crossing the road the case vehicle is turning into should be detected to avoid a crash in front of the vehicle. Also VRU's at the right side close to the case vehicle should be detected to avoid a crash when the truck moves laterally to the right in the turn. Both these areas may contain blind spots and are out of direct view for the driver.

The subsequent Figure 9 shows the principle optical design of the MiniFaros laser scanner based on a coaxial optical concept. The laser diode emits short laser pulses towards the front side of the rotating MEMS mirror. The mirror deflects the laser beam by 30 ° back into the sender lens. This omni-directional lens guides the beam such that it leaves the scanner horizontally. The backscattered pulse is entering the receiver lens and guided to the MEMS mirror and reflected into the avalanche photo diode (APD).

The mayor requirements for the laser scanner are a range of 80 m, a field of view of 250° and an update frequency of 25 Hz.

APD
laser diode
lens
receiver lens
MEMS mirror
sender lens
laser diode

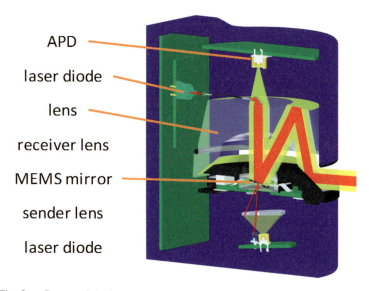

Fig. 9.   Proposed design concept

4.   Conclusions

The project "Low cost miniature laser scanner for environmental perception", MiniFaros, is a sensor development project aimed to significantly increase the penetration of advanced driver assistance systems, ADAS, in the automotive market. A typical top-down approach derives from accidentology to relevant scenarios to selected scenarios to applications finally the laser scanner specifications.

References

[1]   User needs and operational requirements for MiniFaros assistance system. MiniFaros Deliverable D3.1.

Kay Fuerstenberg, Florian Ahlers
Sick AG
Merkurring 20
22143 Hamburg
Germany
kay.fuerstenberg@sick.de
florian.ahlers@sick.de

Keywords: laser scanner, ADAS, MEMS, omnidirectional mirror, EC project, MiniFaros

MEMS Mirror for Low Cost Laser Scanners

U. Hofmann, J. Janes, Fraunhofer ISIT

Abstract

The concept and design of a low cost two-axes MEMS scanning mirror with an aperture size of 7 millimetres for a compact automotive LIDAR sensor is presented. Hermetic vacuum encapsulation and stacked vertical comb drives are the key features to enable a large tilt angle of 15 degrees. A tripod MEMS mirror design provides an advantageous ratio of mirror aperture and chip size and allows circular laser scanning.

1   Introduction

LIDAR sensors are becoming increasingly interesting for the realization and improvement of driver assistance systems like pre-crash safety systems, intersection assistant, lane change assistant, blind spot assistant, parking assistant or traffic jam assistant. A wide angular range and high angular resolution are key-features that scanning LIDAR systems offer. Existing scanning LIDAR systems use bulky servo motors for rotation of a large aperture scanning mirror making it difficult to demonstrate the required sensor dimensions and sensor costs for a series automotive products. But cost reduction and a higher level of miniaturization seem to be possible by introduction of MEMS technology. This paper describes the concept and the design of a low cost two-axis MEMS scanning mirror that aims at replacing the bulky and expensive conventional laser scanner in an automotive LIDAR sensor application.

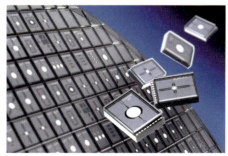

Fig. 1.   Hermetically vacuum packaged two-axis MEMS scanning mirrors fabricated on 8 inch silicon wafers

2   Basic Optical Concept

The key feature of the low-cost LIDAR sensor is an omnidirectional lens that integrates several reflective and refractive functions within one single component (Fig. 2). Omnidirectional scanning is achieved by first collimating the divergent laser beam by passing the refractive centre area of the omnidirectional lens. The collimated beam then impinges on a 2-axis MEMS scanning mirror. The tilted mirror reflects the beam back to propagate through the lens again. After passing two internal reflections at two reflective lens facets the beam exits the omnidirectional lens almost perpendicularly to the optical axis of the incoming divergent laser beam. According to the cylindrical symmetry of the overall configuration the laser beam can be scanned within the whole range of 360 degrees. The optical concept requires a two-axis MEMS scanning mirror which performs a circular scan at a constant rotational tilt angle of 15 degrees resulting in a cylinder symmetric optical deflection of 30 degrees. In order to enable a long measurement range of up to 80 metres the optical configuration requires a mirror diameter of 7mm.

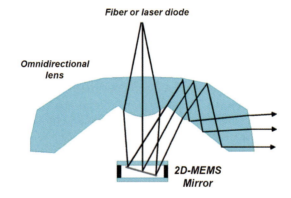

Fig. 2.   The optical concept of the LIDAR sensor is based on a circular scanning MEMS mirror and an omnidirectional lens

3   MEMS Mirror Design

MEMS scanning mirrors have been used in many different applications as for instance barcode scanners, laser printers, endoscopes, laser scanning microscopes or laser projection displays [1]. Typically MEMS mirrors have a mirror aperture size within the range of 0.5 to 2 millimetres. There are two major reasons for the limitation of MEMS mirrors to such small dimensions: Firstly, static and dynamic mirror deformations rapidly increase with increasing mirror diameter and secondly, the very low driving forces of MEMS actuators usually do not allow a reasonable tilt angle of high inertia mirrors. Hence, to design and fabricate a 2D-MEMS scanning mirror with an outstanding mirror size of 7 mm and a large mechanical tilt angle of +/-15 degrees is quite a challenging task.

3.1 Static and Dynamic Mirror Deformation

The optical conception of the LIDAR sensor requires that deformation of the MEMS mirror plate does not exceed +/-500 nanometres. Deformations can be caused by stress gradients within the layers which the mirror is being made of. Typically the uppermost reflective layer introduces mechanical stress that deforms the mirror to some extent. But more often deformation is predominantly caused by the MEMS mirror dynamics. The dynamic mirror deformation is known to scale proportional to the fifth power of mirror diameter [2]:

$$mirror\ deformation \propto \frac{D^5 f^2 \theta}{t^2} \qquad (1)$$

(D = mirror diameter, f = tilting frequency, θ = tilt angle, t = mirror thickness)

This scaling law indicates that to keep the deformation of a mirror with an aperture of 7 millimetres and a tilt angle of 15 degrees sufficiently low needs to correctly adjust the thickness of the mirror. For a more detailed investigation on how the mirror geometry may effect the dynamic mirror deformation finite element analysis (FEA) was carried out. Three different types of mirrors were simulated: 1) a mirror plate having a standard thickness of 80 microns (typical MEMS device layer thickness), 2) a mirror plate identical to first type but additionally reinforced by a 500 micron thick and 200 microns wide stiffening ring underneath the mirror plate, 3) a solid mirror plate with a thickness of 580 microns. For each type of mirror the simulation of mirror deformation was performed for four different diameters (Fig. 3).

The FEA showed that a 7mm-mirror with a standard thickness of 80 microns would experience unacceptably large deformations exceeding +/-6 microns. Considerable reduction of mirror deformation to only +/-1.2 microns can be achieved by a narrow but 500 microns thick reinforcement ring underneath the mirror. Finally a solid mirror plate with a thickness of 580 microns achieved the best result and showed a minimized mirror deformation of only +/-0.2 microns. Thus, further design assessments and simulations only considered the two reinforced mirror types.

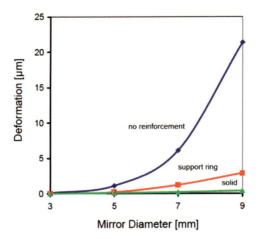

Fig. 3.   Calculated mirror deformation versus mirror diameter for three dif-
          ferent mirror geometries: mirror without reinforcement structure
          80 μm thick, mirror with 500 μm thick stiffening ring underneath
          and solid mirror with thickness of 580μm

3.2   Driving Concept and Fabrication Process

In principle electromagnetic actuation would enable to achieve the highest
driving forces and hence would be the first choice for actuation of such a
high inertia MEMS mirror. But the attractiveness is lowered by the fact that it
requires mounting of large permanent magnets on chip level resulting in a too
large and too expensive scanning device. A compact and cost effective solution is an electrostatically driven MEMS mirror since the whole device can be
produced completely on wafer level including hermetic packaging [3]. Figure 4
shows a two-axes MEMS scanning mirror electrostatically actuated by stacked
vertical comb drives. To drive a large MEMS mirror with an aperture size of
7millimetres to the required large tilt angles of +/-15 degrees it is necessary to
apply resonant actuation because it allows to achieve higher oscillation ampli-
tudes. However, if the MEMS mirror works in standard atmosphere damping
by gas molecules is such high that even resonant actuation is not sufficient to
achieve the required scan angles. To meet the requirements of large mirror size
and large tilt angle it is necessary to additionally minimize damping. This can
be achieved by packaging the 2D-MEMS scanning on wafer level in a miniature
vacuum environment (see Fig. 1). This allows the MEMS mirror to accumulate
driving energy over many thousand oscillation cycles. Electrostatically driven
MEMS mirrors with Q-factors as high as 145,000 have already been demonstrated [4]. Hence, the low-cost LIDAR MEMS scanning mirror will be fabri-

cated in a dual layer thick poly-silicon process. Deposition of thick poly-silicon layers, chemical mechanical polishing, photolithography and deep reactive ion etching are the fundamental processing steps within the MEMS mirror fabrication sequence. Wafer bonding techniques will be applied to permanently protect each MEMS mirror against contamination by particles, fluids or gases. A titanium getter layer will be integrated into each MEMS scanner cavity in order to achieve a permanent miniature vacuum environment.

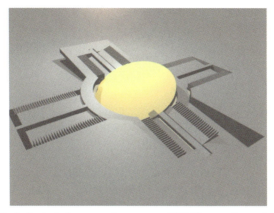

Fig. 4. Typical gimbal-mounted two-axes MEMS scanning mirror electro-statically driven by stacked vertical comb drive

3.3 Suspension concept

As shown in Fig. 4 the standard design to allow a MEMS mirror to scan a laser beam in two dimensions is a gimbal mounted device. But the optical concept of the targeted low-cost LIDAR sensor requires a circular scan trajectory and the MEMS mirror has to provide two perpendicular scan axes that have identical scan frequency. Practically, this is difficult to achieve using a gimbal mounted mirror design. For that reason a completely different design was chosen which eliminates the need for an outer gimbal frame. Instead of suspending the mirror by two torsional beams the mirror plate is movably kept by three long and circular bending beams (see Fig. 5). This allows achieving an advantageous ratio of mirror diameter and chip size which is an important factor for a low cost scanner. Because of a considerably lower total mass with respect to a gimbal mirror design such a tripod design shows higher robustness. Finite element analysis has shown that mechanical stress in the bending beams can be kept sufficiently low to enable the required tilt angle of 15 degrees. Regardless of the three beams which are spatially separated by angles of 120 degree the mirror builds two perpendicular tilt axes (two eigenmodes) that have almost identical resonant frequencies. In comparison with a gimbal mounted mirror design the tripod approach allows to achieve an advantageous eigenmode spectrum showing a large gap in frequency between desired tilting modes and disadvantageous parasitic eigenmodes. Different variants of such a tripod MEMS mirror design will be fabricated covering a range of scan frequency of 600Hz to 1.6kHz. This scan frequency depends on the stiffness of the three suspensions and on the moment of inertia which is different for a solid reinforced mirror and for a mirror reinforced by only a ring. As a result the whole 360 degree scenery in principle can be scanned at a rate of 600Hz or higher.

Fig. 5.   Tripod MEMS mirror design. Deformations are minimized by stiffening rings underneath the mirror plate.

Acknowledgement

MiniFaros is part of the 7th Framework Programme, funded by the European Commission. The partners thank the European Commission for supporting the work of this project.

References

[1]   Hofmann U., et al., Wafer-level vacuum packaged micro-scanning mirrors for compact laser projection displays, Proc. SPIE Vol. 6887, 2008.

[2]   Brosens P., Dynamic mirror distortions in optical scanning, Applied Optics, vol. 11, p. 2988-2989, 1972.

[3]   Oldsen M., et al., A Novel Fabrication Technology for Waferlevel Vacuum Packaged Microscanning Mirrors, Proc. 9th Electronics Packaging Technology Conference, Singapore, 2007.

[4]   Hofmann U., et al., MEMS scanning laser projection based on high-Q vacuum packaged 2D-resonators, Proc. SPIE 2011.

Ulrich Hofmann, Joachim Janes
Fraunhofer Institute for Silicon Technology ISIT
Fraunhofer Strasse 1
25524 Itzehoe
Germany
ulrich.hofmann@isit.fraunhofer.de
joachim.janes@isit.fraunhofer.de

Keywords: MEMS, mirror, LIDAR, laser, circular scan, electrostatic, vertical comb drives, low-cost, hermetic package, vacuum

Omnidirectional Lenses for Low Cost Laser Scanners

M. Aikio, VTT Technical Research Center of Finland

Abstract

There is a need for small sensors that can provide 360° field of views in the intelligent vehicle applications. The usual technique has been to use a catadioptric system where a conical shaped mirror is placed in front of a camera, providing 360° horizontal field of view and in the order of tens of degrees of vertical view. The downside of these kinds of systems has been their size, usually ranging around 20 centimetres. A so-called omnidirectional lens can fold the optical path inside the lens decreasing the volume requirements considerably, while still providing comparative optical performance. In this work, two different omnidirectional lens systems are presented, the more common type of this lens images a whole surrounding scenery to an image sensor, providing instant 360° field of view. The other lens can select a known position from the 360° scenery, and provide an undistorted image of it. The other application for this type of lens is a laser scanner that necessitates direction selectivity.

1   Short Survey of Omnidirectional Vision Sensors

Omnidirectional vision systems are not a new invention; the usual approach to expand the field of view of a known camera lens has been to place a conical or hyperbolical mirror in front of the camera [1-3]. Outside of scientific publications, Olympus has several press releases dating from 2008 that show a combined camera and a lens comprising of several refractive and reflective surfaces to provide omnidirectional vision, but it is unknown for the author if this system is currently on the markets. The author would like to point out that there is some engineering or scientific interest behind the 360° vision systems, evidenced by frequent surveys to omnidirectional vision systems [4-5] and omnidirectional workshops (Omnivis) in conferences like ICCV 2005 and ECCV 2004.

For the surprise of the author of this paper, the exact dimensions of the sensors using a conical mirror are usually not mentioned in the publications, but their field of views are. The other interesting note is that the when such a catadioptric component is placed in front of a camera lens, the F-number of the camera used in the tests and measurements is often not mentioned in the texts. The importance of the F-number, or the light gathering capability, relates to the achievable frame rate in varying illumination conditions if the sensor is operating in a moving system. The reason behind this is likely related to the optical design procedure of the conical mirrors and the conditions placed on the camera lens during the design process: more often than not, the camera lens aperture is described as a pinhole. There is no mention of the prices of hyperbolic or, in general, conical mirrors that are used in the systems.

The other standard technique is to use a fish-eye lens in front of a camera, but these systems usually suffer from relatively low amount of information content at the horizontal plane - this region is more useful in vehicle applications - while a large fraction of the image is used by what is directly above or below the fish-eye lens. In addition, it should be noted that when using a fish-eye lens in daylight conditions and when the lens is pointed upwards, the sun is very likely in the field of view of the sensor, possibly complicating the exposure control.

There are several patents relating to the omnidirectional lens systems presented in this paper [6-10], especially Ehud Gal et al. describe several omnidirectional vision systems that are in principle, the same as the first case presented in this paper. The other example of similar lens is shown in [7], where an omnidirectional lens is used to track incoming projectiles. The second case presented in this paper, we believe, is unique and has not been published formerly.

2   Developed Omnidirectional Lens Systems

In this paper, we present two different omnidirectional lens systems, the first one is similar to former work as discussed earlier and provides an image of the surrounding scenery for the image sensor. The second type of an omnidirectional lens uses a beam steering mirror in order to select the scanning angle, and it could be used to get an undistorted image from a known direction. The other application of this lens is laser scanners, where direction selectivity is very important.

2.1 Omnidirectional Lens with an Image Sensor

Originally initiated and developed in 2006 for mobile conferencing applications, the omnidirectional lens was designed to be used with a mobile phone camera, providing overall small vision system module. Figure 1 represents the working principle of the omnidirectional lens, while a picture of the lens module is shown in Figure 2. A picture taken with this lens is represented in Figure 3, and shows typical distortion patterns for this type of lens. The lens was designed in VTT, and manufactured by diamond turning by a Danish company Kaleido Technologies.

The resulting omnidirectional lens diameter is roughly 33 mm, and the height is roughly 25 mm, including the mobile phone camera. The vertical field of view is 30°, and 360° horizontal, without any shadowing support mechanics. The omnidirectional lens tilts and preserves the collimation state of the incoming beam, and the mobile phone camera itself is used for focusing. The illumination levels are limited by the camera lens aperture stop, the design aperture was F/2.8. The physical aperture diameter of the camera in the design is 1.75 mm, and the focal length was 4.95 mm and full field of view is 40°. The diameter of the collimated pencil of rays that strikes the cylindrical part of the omnidirectional lens is averagely 0.5 mm per field point, when the target is assumed to be far.

The vertical field of view is designed for video conferencing, so that the lens system does not see below horizontal plane, as the phone was assumed to be located on a table. The 30° vertical FOV allows for photographing persons around the table, if they are sitting about 1.5 meters from the mobile phone. The lens material is ZEONEX E48R, which is a plastic that has a good environmental resistance. The lens has been designed injection molding process in mind, which enables low cost manufacturing of optical components. In orders of 100 000 pieces, the price of a single lens is usually counted in cents.

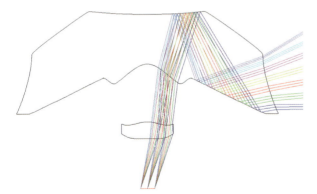

Fig. 1. The working principle of the omnidirectional lens. Under the main lens there is a wavefront correcting element, and only the aperture stop of the camera lens is shown below.

There is an additional wavefront correcting element in the omnidirectional lens system as shown in Figure 1, it is placed under the omnidirectional lens and just above the mobile phone camera aperture. The role of this lens is to improve the incoming wavefront so that a better quality image can be obtained and it is not necessary for all applications, depending on the needed image quality and camera optics. In this case, the mobile phone camera had a sensor of 1280 x 960 pixels, and it was determined that in order to keep the drawing capability of the mobile phone camera unaffected by the omnidirectional lens, and additional lens was required.

Fig. 2.   Omnidirectional lens placed on top of a mobile phone camera. The system size is easily understood from the image.

Our experiences with this kind of combination of a mobile phone camera (or similar) and the omnidirectional lens are that this configuration will allow for a small surround vision sensor size. There are some issues when directly attaching the omnidirectional lens on top of a mobile phone camera with intelligent exposure control; even inside a building, it is common to find a window or other brightly lit area that will dominate the image exposure, leaving other regions slightly under-exposed. When considering the applications of this type of sensor for example in a vehicle, the important consideration is that the sun is surprisingly often in the direct field of view (Fig. 3) of the camera in the northern latitudes and it is our recommendation that this effect should be considered in the design phase. Software correction of this could be possible, but during the project, this was judged to be outside of the scope of the omnidirectional lens research.

Third important consideration is the sensor type itself. If the sensor is constructed with a mobile phone camera or similar, and the application would be a vehicle moving with higher speeds, it is important to select a sensor that does not have a rolling shutter. Otherwise as with normal cameras, there will be additional image distortion caused by the shutter.

Fig. 3.　A picture captured by the omnidirectional lens, seen directly from the camera before the polar transformation or other image correction algorithms. The sky is often slightly overexposed in pictures taken with this camera.

2.2　Omnidirectional lens with direction selectivity

The development history of this lens is related to laser scanner applications, and the general objective of the current Minifaros-project is to replace a large rotating mirror from laser scanners with a MEMS mirror. Instead of imaging a whole scenery around the lens, a rotating mirror is used to select a portion of the scenery to be imaged on the sensor – or to be measured with a laser scanner. Without this property, the laser scanner would not be possible. This kind of lens is new to author's knowledge, and no prior art work has been published of it. The working principle of the lens is shown in Figure 4, and one manufactured lens is shown in Figure 5.

The lens diameter is roughly 50 mm and the height is roughly 25 mm. The outgoing beam is slightly elliptical, resulting in divergences of 30 mrad x 22 mrad with a circular receiver of 200 μm in diameter. The beam area on the cylindrical surface is 23.8 mm^2, which gives the physical limit for power that can be collected on the receiver. When the diameter of the source is less than the diameter of the receiver, the resulting divergence is improved, which allows slightly better resolution. The lens can be used both in biaxial and coaxial configurations, depending on the needs of the application.

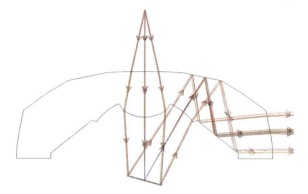

Fig. 4.   A sketch of an omnidirectional lens that has a beam direction capability. The tilted plane below the lens is a beam steering mirror, that is used to deflect and rotate the scanning position. This provides a 360° scanning capability.

Fig. 5.   A manufactured omnidirectional lens which is used in conjunction with a beam steering mirror.

A biaxial laser scanner consisting of two lenses as shown in Figure 5 was constructed, and the performance was evaluated. The divergence of the sensor was 30 milliradians with a detector of diameter 200 µm. The signal to noise ratio allowed the usage of the sensor up to 10 metres, with a black diffuse target. Expanding the measurement distance from this is one of the objectives in Minifaros project.

3   Importance of Omnidirectional Sensor

Omnidirectional vision and sensor systems are important in autonomous vehicle operation if the amount of sensors needs to be reduced. By using a large field of view sensor, there is no need to have multiple sensors in a vehicle. However, one constraint on using them has been the size, manufacturing tolerances and the price of the resulting system.

In this work, we have presented two separate omnidirectional vision systems that are small and easier to install for vehicle and robot applications; the first one uses a mobile phone camera to reduce the size of the omnidirectional vision system. The second lens is used in laser scanner application with a rotating beam steering mirror. This type of omnidirectional lens would also allow imaging of the surrounding scenery without distortion if multiple exposures are taken and the avalanche photo diode is replaced with a small image sensor.

The second important factor to consider is the price of the sensor and related optics. Because the omnidirectional lenses presented in this work are roughly 40 to 50 mm in diameter and are made of plastic to allow for easier serial production of this type of optics. In serial productions when the production volume approaches hundreds of thousands of pieces per year, the price for a single omnidirectional lens is around several cents. In Minifaros project, the omnidirectional lens is used in a laser scanner application (LIDAR) to prevent and mitigate the consequences of vehicle accidents.

Acknowledgements

MiniFaros is part of the 7th Framework Programme, funded by the European Commission. The partners thank the European Commission for supporting the work of this project.

References

[1]   Hrabar, S., Sukhatme, G., Omnidirectional Vision for an Autonomous Helicopter, Proceedings of IEEE International Conference on Robotics and Automation, 3602-3609, 2004.

[2]   de Souza, G., Motta, J., Simulation of an omnidirectional catadioptric vision system with hyperbolic double lobed mirror for robot navigation, ABCM Symposium Series in Mechatronics, Volume 3, 613-622, 2008.

[3] Lima, A., et al., Omni-directional catadioptric vision for soccer robots, Robotics and Autonomous Systems, Volume 36, 87-102, 2001.

[4] Fernando, C. et al, Catadioptric Vision Systems: Survey, Proceedings of the Thirty-Seventh Southeastern Symposium on System Theory, 443-446, 2005.

[5] Yagi, Y., Yokoya, N., Omnidirectional Vision: A Survey On Sensors and Applications, Transactions of Information Processing Society of Japan, Volume 42, 1-18, 2001.

[6] Gal, E. et al, Self-contained panoramic or spherical imaging device, USP 7643052 B2, 2010.

[7] Agurok, I. et al, Passive Electro-Optical Tracker, USP 0278387 A1, 2010.

[8] Ge, Z., Mohcitate, S., Omnidirectional Imaging Apparatus, USP 0309957 A1, 2009.

[9] Gal, E., et al, Omni-directional imaging and illumination assembly, USP 7570437 B2, 2009.

[10] Ito, M., Murayama, O., Omni directional photographing device, USP 0002969 A1, 2008.

Mika Aikio
VTT Technical Research Center of Finland
Kaitoväylä 1, P.O.Box 1100
90571 Oulu
Finland
mika.aikio@vtt.fi

Keywords: omnidirectional vision, catadioptric lens, laser scanner

PSI5 Interface for Ultra Compact Inertial Sensor Clusters

M. Reimann, B. Rogge, J. Seidel, F. Hirschauer, J. Schier, T. Schrimpf, A. Hepp, Robert Bosch GmbH

Abstract

A new interface for inertial sensor clusters is presented. The PSI5 interface offers several advantages and especially reduces the technical complexity to a minimum. Different configurations offer a solution for all VDM (vehicle dynamics management) demands.

1 Inertial Sensor Clusters and Applications

As the MM3 sensor cluster [1] by now counts over 30 million devices in field service, the inertial sensor cluster has found its market in nearly all automobile VDM applications. It provides the signals of angular rate und acceleration to the connected systems while several configurations and measurement axes are available. For example, the ESP® applications are provided with yaw rate (vehicles z axis) and lateral acceleration signals (y axis), see table 1.

VDM Application	Ω_x	Ω_y	Ω_z	a_x	a_y	a_z	Sensor Cluster configuration
ESP®			X		X		MM3.8
HHC				X			MM3.8
ROSE	X					X	MM3.10
Damper dynamics	X	X				X	DCU

Table 1. Examples of sensor cluster configurations for different applications (vehicle dynamics management VDM, angular rate Ω, acceleration a, electronic stability programm ESP®, hill hold control HHC, rollover sensing ROSE)

So far, the specifications of different vehicle applications could be unified and all the systems were equipped with the same sensing elements. This trend continued with the next generation MM5, which entered the market in 2010 using the progress in MEMS and ASIC processes to obtain improved performance with significantly reduced size and costs. For the first time sensing structures of angular rate and two channels of acceleration could be packaged in one single sensor module for automotive safety applications. Compared with its predecessor MM3 the new cluster is up to 60 percent smaller, which significantly simplifies integration into the vehicle. It is compatible with the current variant, and can therefore be used in existing series projects without any complex or expensive adaption.

Fig. 1.   Sensor cluster MM5. The sensing element implies measurement channels for angular rate and two accelerations, integrated safety controller and CAN interface.

Despite its highly integrated design, the signal quality continues to meet the very high standards for dynamics and precision that have to be fulfilled; at some point even better performance is achieved. For the measurements the sensor cluster uses measuring elements in surface micromechanics. The yaw rate sensor's measuring element works according to the Coriolis principle, i.e., it utilizes the inertia force of an oscillating mass in a rotating system. Due to the high resonance frequency of 15 kHz and the closed drive and evaluation unit, the yaw rate element is very insensitive to mechanical interference. Acceleration is measured on the basis of the seismic mass capacitive change in the micromechanical structures.

The safety and monitoring concept is integrated in the sensor module as well. It also covers the entire signal chain: the measuring element, the evaluation electronics, and the microcontroller. The sensor signals can be issued via a standard CAN interface with a variable signal update rate, which, for the first time, is integrated in the sensor module. Therefore, they are also available for all other functions and systems in the vehicle.

As shown above, digital integration and miniaturization are the key drivers to provide a large field of sensing functions. Consequently, the authors want to present a new interface for future sensor generations in the context of inertial sensor integration trends into high performance ECUs.

2   PSI5 interface

In comparison to the cluster generations with CAN interface, the change over to PSI5 unites several advantages. Providing a significantly smaller and flat housing the future generations will be integrated easily, especially in the space critical location of the vehicles' centre of gravity (e.g. under the cup holder above the middle tunnel).

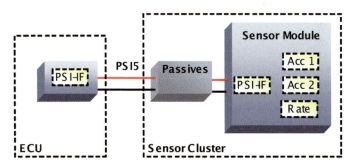

Fig. 2.   PSI5 architecture for a peripheral sensor cluster.

Heart and brain of the devices are sensor modules with integrated channels for acceleration and angular rate. Several different hardware filter options and modularity up to 6D with optional redundancy are available. Enhanced compactness for future cluster generations will be enabled by innovation in the field of sensor networking within the safety domain: the PSI5 bus (Peripheral Sensor Interface). It is a common and known standard in today's passive safety systems with a two wire connection including power supply and communication interface. The PSI5 protocol will be enhanced to a new standard to fulfil the data rate requirements and modularity of this new application in the chassis domain. A first draft of this new specification (V2.0) was proposed to the PSI5 consortium in January 2011 (see [2]).

2.1 Physical Interface Basics

PSI5 is communication is based on the principle of current modulation. Two wires (GND and supply) connect the sensor with the PSI5 receiver inside the ECU. Via these two lines the sensor is supplied and communicates with the ECU.

The static condition is defined by the quiescent current of the sensor. When data has to be transmitted from sensor to ECU, a current source inside of the sensors PSI5 controller is switched to modulate the current consumption. These current changes will be detected by the transceiver through a shunt resistor. This resistor translates the current signal into a voltage signal. Two advantages of the interface are the high sending current of 26 mA which results into a high SNR, so that a high immunity to EMC events is given. Furthermore, relatively high capacitive load in the range of 30 to 100 nF enables the suppression of high frequency noise. The bits are encoded with Manchester code. Each bit is coded by a rising (0) or falling edge (1) of the current level. The following figure shows the principle of the Manchester coding.

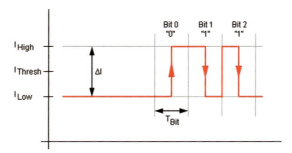

Fig. 3.   Physical principle of Manchester coding, rising edges representing a logical "0", falling edges a logical "1", [3].

2.1 Data Link Layer

The new PSI V2.0 standard permits that each data packet consists of 2 start bits and of an additional variable number from 10 to 28 data bits. For data integrity a parity bit or a 3-bit CRC-checksum are transmitted at the end of each data packet. The gap time (time between two data packages) is at least the maximal length of one bit between different data packets.

For application in sensor clusters, 20 bit wide data words for data transmission are planned (see Fig. 4) to ensure the data rates and resolutions which are already obtained today via CAN.

The channel bits encode the type and orientation of the signal data, the signal status shows the validity of the signal and the 16 bit data field contains values of the signal data.

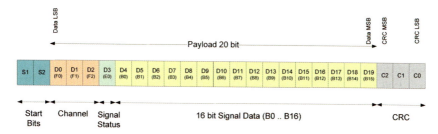

Fig. 4. New 20 bit data word structure in PSI5 V2.0 standard.

During initialization the sensor cluster performs internal self-tests and transmits a special init sequence for every signal channel which consists for instance of manufacturer code, manufacturing date, protocol type, and a serial number. After successful initialization the sensor cyclically sends its sensor data. In case of error detection the signal status marks the signal data as invalid.

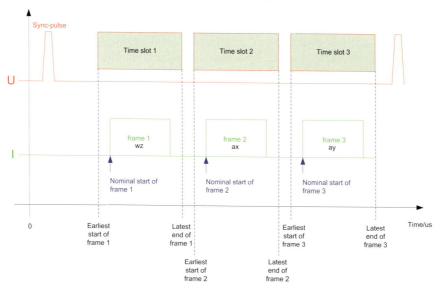

Fig. 5. Timing scheme for synchronous PSI5 communication mode.

The PSI5 sensor cluster offers different operating modes described in the new substandard "Vehicle Dynamics Control V2.0". These modes are called synchronous and asynchronous. In asynchronous mode the sensor works independently of the ECU. The sensor transmits continuously 3 frames every 900 µs. After having performed the initialization the sensor continuously transmits 3 signal data frames to the ECU every 900 µs. The receiver synchronizes with the bitrate of the sensor through the two start bits. This recommended operation mode for chassis systems and VDM functions can be found in the substandard description as A2OCRC 300/1L.

In synchronous mode the sensor is triggered by the ECU receiver through a sync-pulse sent every 500 µs. This sync-pulse is detected by the sensor. The sensor sends 3 data frames after detection of this pulse within 500 us. The absolute start times for sending the data frames can be configured inside of the sensor individually. This operation mode is described in the substandard as P2OCRC 500/3H (see fig. 5).

Beside the synchronous and asynchronous mode a special bidirectional communication mode called tooth gap method is supported for sensor programming and diagnostic purpose.

3   System Architecture Considerations

During the last few years several automotive architecture options for inertial sensors have emerged. Inertial sensors are integrated in different ECUs in the passive and active safety sector and partly replace the external placement of the sensor cluster. However, with the integration of the inertial sensors in ECUs the specific requirements of inertial sensors in VDM applications have now been added to the mechanical needs of an already complex ECU design.

Robustness to vibrations is a key performance parameter for inertial sensors in automotive applications. Today, the sensor clusters fulfil this requirement by using a vibration-optimized housing in combination with robust sensing elements. The housing optimization is achieved with FEM calculations and laser vibrometer based analysis. In comparison to other forms of integration with high Q PCB resonances and therefore high impact on sensor signal quality the compact design enables low Q resonances with best signal performance even in roughest automotive conditions. Additionally, no variances in the mechanical housing of the sensor cluster enable a profound and optimized design for vibration robustness and thus a plug-and-play solution for vehicle application.Even to this mechanical question the PSI5 interface adds its benefit: The PSI5 power supply architecture enables the sensor cluster to omit bulky populated SMD components like electrolyte capacitors. Bulky components can be stimulated by external vehicle vibration and produce rattling noises which could disturb the sensor signals.

With the configurability and scalability of the PSI5 interface, the next generation of sensor clusters will be the right answer for the growing demands for signal quality und flexibility in active and passive automotive safety systems.

References

[1]   Schier, J., Willig, R., New Inertial Sensor Cluster for Vehicle Dynamics Systems, Advanced Microsystems for Automotive Applications 2005, pp. 269, 2005.
[2]   www.psi5.org
[3]   Adam, B., Brandt, T., Henn, R., Reiss, S., Lang, M., Ohl, C., A New Micromechanical Pressure Sensor for Automotive Airbag Applications, Advanced Microsystems for Automotive Applications 2008, pp. 259, 2008.

Mathias Reimann, Berthold Rogge, Jana Seidel, Frank Hirschauer
Johannes Schier
Robert Bosch GmbH Abstatt
Postfach 13 55
74003 Heilbronn
Germany
mathias.reimann@de.bosch.com
berthold.rogge@de.bosch.com
jana.seidel@de.bosch.com
frank.hirschauer@de.bosch.com
johannes.schier@de.bosch.com

Thomas Schrimpf, Aline Hepp
Robert Bosch GmbH Reutlingen
Tübinger Str. 123
72762 Reutlingen
Germany
thomas.schrimpf@de.bosch.com
aline.hepp@de.bosch.com

Keywords: sensor, inertial, cluster, angular rate, acceleration, PSI5, VDM, ESP

Networked Vehicle

A User-Centric Approach for Efficient Daily Mobility Planning in E-Vehicle Infrastructure Networks

N. Hoch, B. Werther, H. P. Bensler, Volkswagen AG
N. Masuch,, M. Lützenberger, A. Heßler, S. Albayrak, TU Berlin
R.Y. Siegwart, ETH Zürich

Abstract

The next generation of e-vehicles will be assessed on their ability to master the challenges posed by urbanization, resource restrictions and increasingly flexible and diverse user demands. Presumably, the vehicle-user-infrastructure network will become too dynamic and complex for an individual user to fathom without some sort of automated assistance. This paper proposes a user-centric, constraint-based, in-vehicle travel planning system, which schedules a daily travel plan of a user by exploiting knowledge about current and future states of the vehicle-user-infrastructure network. The system is implemented in an agent-based framework and is evaluated by a traffic simulator that additionally incorporates parking and charging lots together with an availability monitoring and booking service. The benefit of the planning system is assessed in a traffic simulation, where vehicles with and without a planning system compete for available resources. It can be shown that proactive conflict management and resource scheduling can reduce parking search time and overall travel time. Furthermore it can be shown that the adherence to schedule can be significantly improved.

1 Introduction

The world population is predicted to reach 8 Billion by 2030 with 60.8% expected to live in cities (http://esa.un.org/unpd/wup/index.htm). However, urbanization has different facets in different parts of the world. E-mobility solutions will have to innovate alongside the diversely emerging city characteristics and similarly diverse mobility requirements. A one-fit-all mobility solution seems highly unlikely. Widely successful mobility solutions will feature adaptable system architectures that are scalable to the size, population density, traffic volume and infrastructure characteristics of the city. The technology will need to be capable of dealing with resource restrictions, such as parking space, energy and time.

Currently, for individual motorized mobility, a user primarily plans its journey based upon experience and intuition or by consulting external services. For well-known destinations the user has a sense of when to depart in order to reach the destination in time. The user also has a sense of how much time to allow for parking and travelling from the car park to the destination. But, under certain conditions the user loses the ability to plan its daily mobility pattern efficiently. This occurs when dealing with unknown environments, disruptive technologies, e.g. electric vehicles, or when the environment becomes too complex for the user to easily understand and predict the consequences of its actions. Gain in travel schedule efficiency from travel planning requires from the user both lengthy information search procedures of future vehicle infrastructure conditions and reasoning upon the available information. Yet, users seem to have reluctance towards high cognitive load and lengthy search, as indicated in e.g. [7], [8]. Users demand simplicity together with flexibility and reliability of mobility without the necessity of planning, experience or reasoning skills. Additionally, environment friendliness and adaptability to the user's preferences are key factors for the sustainable market success of future mobility solutions.

In the face of conflicting requirements from infrastructure, e-vehicles and users, we argue that an automated travel planning system can resolve scheduling conflicts and provide simple and efficient individual mobility to the user. The manual planning effort and driver anxiety can simultaneously be reduced whilst improving the individual travel plan. In Section 2 research questions and objectives are deduced. In Section 3 the prototypical planning system is derived. It tries to exploit prior knowledge of the user's daily calendar in order to proactively optimize the user's travel plan. Moreover, it integrates resource availability into the daily scheduling in order to improve the predictability and efficiency of individual routes. In Section 4 the benefit of the planning system is quantified by simulation, in which unplanned and planned vehicles compete for resources. Finally, a discussion of the results is provided.

2 Problem Statement

Intelligent e-vehicles interact with their surrounding infrastructure components. Information about past, current and future infrastructure states is made available to the e-vehicle through vehicle telematics services. An e-vehicle's route sequence is governed by its driver's appointment sequence as well as the driver's specific preferences. At all times the e-vehicle obeys its technical constraints such as power output or recuperation restrictions. Within the traffic scenario the e-vehicle is competing for resources whilst obeying driver, vehicle and infrastructure hard constraints. A route sequence is feasible if the vehicle can reach each user-defined calendar appointment in time, if each route contains both available and suitable parking and charging lots and if the vehicle's state-of-charge never under-runs a limit energy value of the battery. A route sequence is optimal if battery state of charge, adherence to user schedule and choice of parking and charging lot are optimally matched against the user preferences and if minimal travel time and energy are consumed.

2.1 Objectives

The high-level objective of the herein discussed planning system is to optimally fulfil user preferences whilst raising time efficiency of the daily vehicle travel plan. The following mobility related user preferences are investigated: (a) Reduction in travel time and distance; (b) increase in usable time throughout the day; (c) adherence to user schedule; (d) adherence to user preferences. The benefit of the planning approach is quantified in simulation, where planned and unplanned vehicles compete in an equally allocating traffic scenario. Planning an efficient daily mobility pattern can be interpreted as a multi-criteria optimization problem dealing with different objective functions, such as the maximization of usable working time, the minimization of driving time and the minimization of appointment conflicts.

2.2 Approach

The travel planner automatically transforms a user's daily calendar into a vehicle travel plan. The travel planner schedules both pedestrian and vehicle routes. Pedestrian routes connect the appointment location to the parking lot. Vehicle routes connect parking lots of consecutive but different appointment locations. The route sequence is generated under consideration of the user-specific constraints such as maximum walking distance from the parking place to the appointment location and under consideration of infrastructural constraints such as traffic flow of the street network and parking lot availability. It has been argued that proactive planning has the potential to increase the quality of the daily travel plan whilst reducing the driver's planning effort. Within the scope of this paper, this is achieved by the culmination of three approaches. The type 1 approach analyses the appointment sequence of a user calendar in order to proactively detect conflicts within the sequence. Conflicts occur if, under consideration of the vehicle performance constraints, the time-space motion pattern of the vehicle is infeasible. The type 2 approach reorganizes the appointment sequence so that usable time between appointments is maximized. The type 3 approach integrates proactive resource scheduling into routing.

3 Method

3.1 System Architecture

The mobility system can be conceptually modelled as an ensemble of entities such as a user, a vehicle or a parking lot operator (http://www.ascens-ist.eu/index.php). The system architecture must meet scalability, adaptability and privacy requirements. A centralized approach is not inherently adaptable or scalable; nor does it inherently respect privacy issues. A better option, when it comes to system architecture, is the development of a multi-agent system, which has a decentralized structure. It offers the possibility to define autonomous agents with constrained knowledge, which can be distributed on different devices, such as a smart phone, a vehicle PC or a backend server.

In doing so, the system is both easily scalable and adaptable because of its modular structure. The notion of privacy is also clearly defined since knowledge is distributed and every autonomous agent controls the information it is publishing. Taking an agent-based approach, the software architecture becomes intuitive due to its structure being closely related to the structure of the real environment. Within the scope of this paper the multi-agent framework JIAC V [5] is used for implementation, since it fulfils all of the above stated requirements. Figure 1 shows the system architecture of the prototypical automated planning system. Five different agents are defined within the system, each of them providing different services and fulfilling different tasks: (a) The User Agent: Represents the user perspective. It analyses appointments with regard to location and time. Further, it evaluates whether any appointment and its respective routes are at odds with other tasks. If so, the agent initiates an interaction with the user to resolve the conflict. The User Agent can be seen as the instance that triggers the automated planning process. (b) The Car Agent: Is responsible for route and parking related decision-making. It receives planning requests from the User Agent and initially decides which route and parking lot to choose. (c) The Calendar Agent: Provides access to a user calendar. It offers different services to the other agents, such as the readout, the modification and the insertion of appointments. (d) The CarPark Agent: Manages a set of parking lots and is able to process booking requests. The agent is also able to answer more complex requests, e.g. enquiries for the nearest vacant parking lot for a given position. (e) The Routing Agent: Features a highly detailed and GPS based map and calculates routes for a given start and destination location, time and under consideration of user preferences. Currently, the agent supports shortest and fastest route calculation, while live conjunction and consumption aspects can be considered as well.

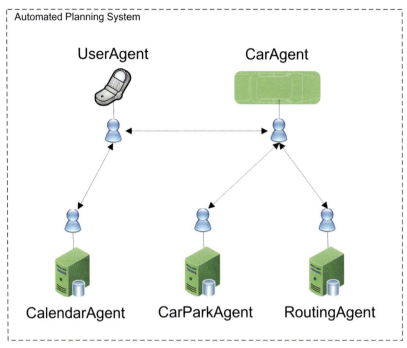

Fig. 1.   System architecture

3.2   Process Workflow

The automated planning process is initiated by the User Agent at certain intervals (see Figure 2). It requests from the Calendar Agent all appointments that have been inserted into the user calendar during the last 24 hours. Using Exchange Web Service Interfaces (http://msdn.microsoft.com/en-us/library/bb204119(v=exchg.140).aspx) the Calendar Agent accesses the user's account and gathers the required information. The User Agent analyses which of the returned appointments have not yet been considered in planning. For each of the unplanned appointments the following procedure is executed. At first the location of the appointment is evaluated. For appointments without location entry the agent assumes the user's office location, which can be defined in the user's preferences. In order to integrate an appointment into the planned appointment sequence the User Agent extracts time and location information about prior and posterior meetings. This initial skeleton of planning is forwarded to the Car Agent.

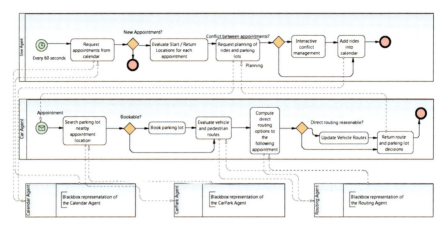

Fig. 2.   Automated planning approach, illustrated by a BPMN diagram

For each appointment location the Car Agent requests all accessible parking lots from the CarPark Agent. The latter evaluates locations of car parks in order to precisely specify the target for the route. Additionally, the CarPark Agent manages and monitors parking space availability over time and offers a booking service in order to reserve particular parking places for a specific time interval. Upon reception of available parking lots the Car Agent requests the Routing Agent. The Routing Agent calculates a route sequence by evaluating the vehicle routes between parking lots of consecutive appointments. The Routing Agent returns time-space information for the routes. Time information is predicted on the basis of the route, traffic and vehicle specific parameters and user handling characteristics. Pedestrian routes are calculated from the parking lot to the meeting location. Eventually all routing and parking lot decisions are returned to the User Agent. The evaluated routes and expected expenditures of time are checked by the User Agent regarding potential conflicts with other appointments within the user's calendar. For example, it might not be possible to get from one appointment to another within the allocated time. The User Agent informs the user about this conflict via e-mail or any mobile device and tries to solve the conflict interactively. Solutions include, among others, shifting the first appointment earlier to allow for sufficient travel time afterwards or completely rescheduling an appointment. The planning process for the appointment sequence is terminated once user interaction is completed. The above-described processes are iteratively executed for each appointment, which finally leads to a complete daily driving schedule. Evaluated journeys and their corresponding expenditure of time are inserted as appointments into the user calendar. The precise routes and the parking lot reservations are stored within the Car Agent.

4   Results

The planning system presented in Section 3 is validated using a traffic simulator [6]. It will be shown that proactive conflict management together with proactive resource scheduling significantly increases time efficiency over the daily travel sequence. Whilst usable time, which is the time available for actually performing work, is increased both the travel time and travel distance can be diminished. The advantage of time and distance savings of planned vehicles over unplanned ones increases when travelling in environments with increasing resource shortage. In the following section, the benefit of the travel planner is quantified exemplarily for varying degrees of parking lot availability as one representative of the class of restricted resources.

4.1 Preconditions

For both planned and unplanned vehicles daily appointment sequences are generated at random for each vehicle. These appointment sequences have a maximum of six appointments and a minimum of two appointments with an average of four appointments per day. Average appointment duration is 60 minutes with minimum and maximum appointment duration of 15 and 120 minutes respectively. Each vehicle maps its appointment sequence to its vehicle travel plan. Planned vehicles perform the mapping by using the herein discussed travel planner. The resulting vehicle travel plan consists of reserved parking places close to the appointment location, pedestrian routes from the parking place to the destination, and vehicle routes that connect the parking places. The resulting appointment sequence is conflict free and vehicle energy levels are planned so that the route sequence is feasible. Unplanned vehicles do not interface the travel planner. Instead, they follow a scheme, which arguably represents the real world behaviour of drivers. The scheme cannot revert to experience or intuition for a particular route or destination. The unplanned vehicle's appointment sequence is transformed into the vehicle travel plan as follows: (a) A route is calculated for the return journey from the home or working location to the appointment location. (b) For each appointment of the daily user calendar the return routes are calculated. (c) Route travel times are extracted. (d) If travel time between consecutive appointments is less than the available time between appointments, the unplanned vehicle executes its travel plan. (e) Otherwise, the unplanned vehicle requests a new appointment sequence. Provided that a feasible travel plan exists, the unplanned vehicle heads towards its first appointment. On arrival, it starts searching for an available parking place. The parking search is performed by travelling expanding circles around the appointment location. The unplanned vehicle will always choose the first available parking place. The simulation records the route travel time and distance of the unplanned vehicle, which includes the walking time and distance. The vehicle considers an appointment as missed if it has unsuccessfully searched for an available parking space before exceeding the start time of the appointment by more than 15% of the appointment duration. When an appointment is missed the appointment related parameters, e.g. walking distance and time, are filtered from the results. The traffic scenario is restricted to a 100km x 120km map section of Berlin. Appointments can be generated for each available street number within the map. Within the map, only the 390 major car parks are considered, whose capacity can be changed virtually in the simulation. The average walking distance from any appointment location to its nearest car park is evaluated to be 1398m. The mean route length between two equally distributed appointment locations within the scenario is simulated to be 18.6km.

4.2 Quantitative Results

The work-travel interval is defined as the time between the departure from work to the first appointment and return to work from the last appointment of the day. The work-travel interval is always less than the day's working hours. In simulation the average work-travel interval averages 335min. An average of 202min out of the 335min is dedicated towards working, with 117min of meeting time and 85min of working time at the office. Hence, the user spends 60.3% of the work-travel interval working, whereas travelling accounts for 39.7%.

Conflict management: If the planner detects potential time conflicts throughout the appointment sequence it attempts to resolve the conflicts either by scheduling direct drives between consecutive appointments or by shifting appointments interactively. Within the simulated scenario the need to reschedule a direct drive in order to resolve a potential appointment conflict arises in average once every second (2.2) appointment sequence.

Integrated resource scheduling: It has been argued in Section 2 that in addition to conflict management, resource scheduling has the potential to improve the quality of the daily travel plan, that is reducing travel time and increasing adherence to both user preferences and appointment schedule. Resource scheduling is measured with respect to single appointments, in contrast to conflict management, which is evaluated against an entire day. The benefit to individual users from resource scheduling is quantified over the ratio of parking space capacity to the number of vehicles in the scenario. The benefit of the planning system increases disproportionately in scenarios of sparse resource availability. Vehicles with a travel planning system are referred to as planned vehicles.

Figure 3 compares the average travel distance per journey for planned and unplanned vehicles. Travel distance is normalized over the scenario's average route length. It can be seen that planned vehicles travel an average of 87% of the scenario's mean route length. Travel distance of planned vehicles is constant over the ratio of parking space capacity to the number of vehicles in the scenario. With the reduction of the relative resource capacity, planned vehicles gain distinct advantage over unplanned vehicles. As shown in Figure 3 resource scheduling can reduce driving distance up to 67%. Towards high relative resource capacity the travel distance for planned and unplanned vehicles converges.

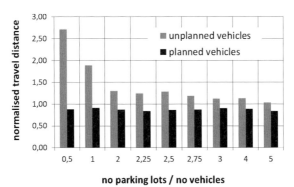

Fig. 3. Average travel distance (per trip) of planned and unplanned vehicles over the ratio of parking space capacity to the number of vehicles in the scenario (relative resource capacity).

Searching for parking spaces generally occurs in areas of high traffic density. Hence, any increase in "parking search distance" disproportionately effects travel time and the trip's average travel speed.

Figure 4 shows the average travel speed of planned vehicles to be 57km/h, which remains constant over relative resource capacity. The travel speed of the unplanned vehicles is on average 4.9km/h lower than that of planned vehicles. Once again planned vehicles become increasingly advantageous with decreasing relative parking space capacity.

In Figure 5 total journey times are discussed. The total journey time is considered to be the sum of the vehicle travel time and the walking time from the parking lot to the appointment location. Journey time is normalized with respect to the simulated journey time at the ratio (parking lots/vehicles) = 5. As can be expected from Figure 3 and Figure 4, the planned vehicles' savings in journey time are especially significant in an environment of low relative parking space capacity.

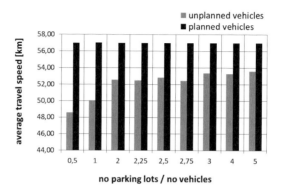

Fig. 4. Average travel speed (per trip) of planned and unplanned vehicles over relative resource capacity.

However, at the ratio of 0.5 and 1 the difference in journey time between planned and unplanned vehicles is unexpectedly small. For example, at relative parking space capacity of 0.5 the journey times of unplanned vehicles do not appear to replicate the significant drop in average travel speed (Figure 4) and the dramatic increase in average travel distance (Figure 3). As has been stated in (Section 4.1), appointments are considered missed if the vehicle does not succeed in finding a parking space before the start time of the appointment is exceeded by 15% of the appointment duration. Vehicles that miss appointments are not counted for evaluation of travel distance, walking time or journey time.

Figure 6 displays the rate of missed appointments of planned and unplanned vehicles with respect to relative parking space capacity. As parking space availability decreases unplanned vehicles increasingly miss appointments. At a ratio of 0.5, unplanned vehicles miss nearly 60% of their appointments and are hence not counted, which explains the unexpectedly low difference in journey time.

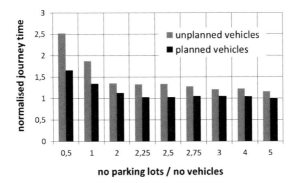

Fig. 5.   Complete journey times (per trip) of planned and unplanned vehicles over relative resource capacity.

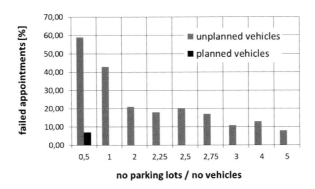

Fig. 6.   Rate of failed appointments (per day) of planned and unplanned
vehicles over relative resource capacity.

5   Conclusion

It could be shown that proactive conflict management resolves, on average,
0.46 conflicts per user, per day. Automatic conflict detection reduces human
planning effort. Interactive conflict resolution respects the user's personal
responsibilities and contributes to the visibility and transparency of the planning system. It could be further shown that resource scheduling reduces the
average travel distance and in particular reduces the parking search distance.
Additionally, average travel speed can be increased whilst overall journey
time can be reduced. Most importantly the rate of missed appointments can

be significantly reduced. The advantage of planned over unplanned vehicles increases in environments of sparse resource availability.

6   Related Work

The paper proposes a software service, which combines appointment scheduling, mobility planning and resource scheduling in order to provide a holistic and automated travel planner for the user.

A comprehensive data collection of both mobility and appointment related user data is provided by [3]. In [8] the usage and the relative attractiveness of appointment planning mechanisms of users such as memory, paper-planner or PDA are investigated. With an increase in the planning related complexity, an increasing number of appointments and greater temporal distance towards appointments, users tend to value the benefits of automated mechanisms such as PDAs. An automated planning system, which combines the increasingly complex and constrained task of appointment scheduling and mobility planning, therefore seems especially beneficial.

As for the time horizon of planning, the effects of pre-trip information on travel behaviour have been investigated by [7]. It has been argued that the imbalance between travel demand and traffic network capacity can only be resolved pre-trip where all decision-making options are still available. Benefits of on-trip information, such as potential savings in travel time due to parking guidance systems (e.g. [2], [9]), have been well studied. Effects are statistically significant, however potentially smaller than the effects of pre-trip information.

The quality of the planning results greatly depends on the information quality that is availability, resolution and certainty of the information. [4] shows exemplarily how information from vehicle telematics units could be aggregated, analysed and distributed efficiently for the special case of parking space information.

Lastly, in the project "eCar" [1] the role of ICT system architectures in electric vehicles has been investigated. In the project's final report an architectural evolution is proposed towards distributed, service-oriented vehicle architectures by 2030.

7 Future Work

The benefit of the planning system has been investigated from the local user perspective. Future work may include an investigation into the global infrastructure perspective, more precisely an evaluation of the effects of proactivity and resource integrated travel planning on infrastructure capacity usage. Simply, how does the penetration level of planned vehicles relate to infrastructural resource availability? Resources under consideration could be parking and charging lots as well as overall energy consumption or overall CO_2 emissions.

Another system extension could focus on charging station availability rather than parking lot availability. In an even more holistic e-mobility planning approach the system could be extended from the pure consideration of charging station availability to energy grid coordination, where e-vehicles consider grid specific constraints throughout the daily route sequence scheduling.

References

[1] The Software Car: Information and Communication Technology (ICT) as an Engine for the Electromobility of the Future, 2011.

[2] Axhausen, K. W. et. al, Effectiveness of the parking guidance information system in frankfurt am main, Traffic Engineering and Control, 35: 304–309, 1994.

[3] Axhausen, K. W. et. al, Observing the rhythms of daily life: A six-week travel diary, Transportation, 29:95–124, 2002.

[4] Caliskan, M. et. al, Predicting parking lot occupancy in vehicular ad hoc networks, In VTC Spring, 277–281, 2007.

[5] Hirsch, B. et. al, Merging Agents and Services — the JIAC Agent Platform, Multi-Agent Programming: Languages, Tools and Applications, 159–185, 2009.

[6] Lützenberger, M. et. al, The BDI driver in a service city, In Proceedings of the 10th International Conference on Autonomous Agents and Multiagent Systems, Taipei, Taiwan, May 2011, to appear.

[7] Polak, J., Jones, P, The acquisition of pre-trip information: a stated preference approach, Transportation, 20, 1993.

[8] Starner, T. E. et. al, Use of mobile appointment scheduling devices., In Proceedings of CHI, ACM Press, 1501–1504, 2004.

[9] Waterson, B.J. et. al, Quantifying the potential savings in travel time resulting from parking guidance systems a simulation case study, JORS, 52:1067–1077, 2001.

Nicklas Hoch, Bernd Werther, Henry Bensler
Volkswagen AG
Berliner Ring 2
38440 Wolfsburg
Germany
nicklas.hoch@volkswagen.de
bernd.werther@volkswagen.de
henry.bensler@volkswagen.de

Nicklas Hoch, Roland Y. Siegwart
ETH Zürich
Rämistrasse 101
8006 Zürich
Switzerland
hochn@ethz.ch
rsiegwart@ethz.ch

Nils Masuch, Marco Lützenberger, Axel Heßler, Sahin Albayrak
Technische Universität Berlin / DAI-Labor
Ernst-Reuter-Platz 7
10623 Berlin
Germany
nils.masuch@dai-labor.de
marco.luetzenberger@dai-labor.de
axel.hessler@dai-labor.de
sahin.albayrak@dai-labor.de

Keywords: Electrified vehicles, networked vehicles, e-mobility, road integration, journey scheduling, proactive resource scheduling, multi-agent systems, autonomous service component ensembles, constraint-based optimization

Using Vehicle Navigation and Journey Information for the Optimal Control of Hybrid and Electric Vehicles

C. Quigley, R. McLaughlin, Warwick Control

Abstract

Environmentally friendly vehicles (electric and hybrid electric) have become essential in today's society. For both it is imperative to optimise the control systems to extend the range of their journeys. Electric Vehicles must have an efficient battery management system to ensure the completion of the journey. Hybrid Vehicle must optimise its control system so that the best balance.of electricity and internal combustion engine is maintained. Modern GPS systems can assist the driver in ascertaining a journey destination and characteristics. This paper explores the possibilities for adopting high-level control strategies for reducing energy usage. Several journey estimation methodologies are described in the form of different control strategies for battery management. It also investigates the possibility to predict the requirements of a journey at its origin, and if an incorrect prediction is made, how this is dealt with.

1 Introduction

Much of the past research has been performed to ascertain the optimal configuration for the environmentally friendly vehicle. Considerations have been made for Zero Emission Vehicles (ZEVs) and Low Emission Vehicles (LEVs) in the form of Electric Vehicles (EV) and Hybrid Electric Vehicles (HEVs). EVs rely on current sophisticated battery technologies with efficient battery management and motor control systems. HEV control is more complicated than for EV since it requires the coordination of Electric Motors (EM) and Internal Combustion Engine (ICE) to propel the vehicle. Two main configuration for the HEV are parallel and series. The parallel drive HEV has both the ICE and EM connected directly to the vehicle transmission and therefore gives rise to at least three possible modes of operation; ZEV mode (propelled solely by the EM and therefore acts as an EV), ICE mode (propelled solely by the ICE) and HEV mode (propelled by a combination of EM and ICE). The series drive HEV is a simpler design with only the EM connected directly to the vehicle transmission and therefore gives rise to only two possible modes of operation; ZEV mode (propelled solely by the EM and therefore acts as an EV) and HEV mode (propelled solely by the EM, but with the ICE used to recharge the battery via a generator). Also, many studies have considered Regenerative Braking for both EV and HEV as a means of recovering lost energy in the braking situation [1, 3, 4, 5, 6, 7, 11].

Navigation information via GPS in modern vehicles is common place as both standard and aftermarket fits and gives very good information on vehicle location and routes. If the precise origin of a journey is known and is combined with the time of day then for many users the likelihood of ascertaining a particular journey and destination is high.

This paper explores the possibilities for adopting high-level control strategies for reducing energy usage in EVs and HEVs using estimated journey information. It is then investigated whether it is possible to predict the requirements of a journey at its origin and secondly, if an incorrect prediction is made, how this is dealt with.

2 HEV and EV Control Strategies Optimised for Journeys

For a long time it has been acknowledged that the usage pattern of a HEV impacts upon its design [2]. Previous studies on HEV control have identified a requirement for a journey specific optimised strategy. The use of adaptive fuzzy decision making in the high level energy management of a parallel HEV was studied [4, 5]. It was suggested and shown that the use of such a technique could be used to adapt the control rules to a particular drive cycle (or even the driver) over about three repetitions of that drive cycle. However this technique required a journey's origin and destination to be predicted and the associated drive cycle for that journey be reasonably repeatable in its characteristics. Another study [6] investigated a number of control strategies for a series HEV. It was suggested that one future improvement to the control of a series HEV would be to incorporate a priori knowledge of a journey's energy requirement and energy regenerative opportunities.

Journey parameter optimised hybrid and electric vehicle control strategies are discussed in this section. The strategies take account for a number of different scenarios that may occur, requiring information to be determined on the journey about to be taken. At the start of a journey, the energy available for its traversal in a HEV is given by the combination of three terms representing the energy in the batteries (or other storage device for that matter), fuel tank for fuelling the ICE and energy that can be recovered through regenerative braking:

$$E_A = E_B * \varepsilon_B + E_R * \varepsilon_R + E_F * \varepsilon_{ICE} \tag{1}$$

For an EV, the third term is not of relevance.

2.1 Minimised Use of ICE in HEV Control

It is usual for the efficiency of the ICE to be considerably worse than the other components. Therefore the goal of journey parameter optimised HEV control is to minimise the use of the ICE. This would be achieved by the use of some simple control rules. In theory, optimised hybrid electric control requires that two parameters be predicted at journey departure. These are:

▶ E_p is the predicted energy requirement for the journey.
▶ ER_p is the predicted energy that can be recovered during the journey.

The predicted energy requirement E_p and predicted energy recoverable ER_p are the parameters that the HEV control is based around. Upon journey departure each of the parameters are predicted and the energy required by each of the powertrain components can be calculated. If:

$$E_p > (E_b * \varepsilon_B + ER_p * \varepsilon_R) \tag{2}$$

then it is assumed the ICE will be required at some point during the journey (i.e. when battery SOC reaches its lower switching threshold).

Figure 1 demonstrates a simplified view of the battery State of Charge (SOC) of a HEV under optimised control. Under this control, it is assumed that the journey's energy requirement has been predicted to be 60 Mjoules, the battery maximum switching threshold is 80% SOC and the minimum switching threshold is 40% SOC. The journey commences with the battery SOC at 80% and the HEV is propelled in ZEV mode. Therefore the battery SOC is depleted down to 40% SOC. At this point, the ICE is started and used to recharge the battery and to supply power to the EM. However, since the journey's energy requirement is known, the battery is only recharged up to 60% SOC. This leaves the battery with only enough energy to complete the journey leaving the battery SOC at 40% (the minimum switching threshold) and with no further assistance from the ICE. Additionally, if it later became apparent that a little more battery discharge was required, the battery could be discharged below its lower switching threshold (so long as it is not depleted low enough to cause irreparable damage). This course of action may be required if minor route deviations or variations in traffic conditions cause an additional energy requirement.

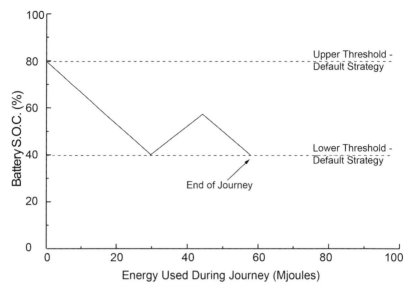

Fig. 1.   A Simplified View of the Battery SOC During a HEV's Journey
Under Optimised Control

The main advantages of this approach to HEV control are:

▸ Reduction of emissions due to reduced ICE start ups. This is particularly
the case if the battery SOC reaches the minimum switching threshold
shortly before the end of the journey. A non-optimised HEV control
strategy would result in an unnecessary ICE start-up.

▸ Possible reduction in net emissions, energy usage and vehicle running costs if domestic electricity supply is used as much as possible to
recharge the vehicle battery.

2.2   Double Lower Limit SOC HEV/EV Control

An extension on the previous control strategy is to allow for two lower limits of
battery SOC. The strategy uses a main lower limit for SOC in all cases except
for when it is established that when the lower SOC limit is reached the vehicle
still has a short way to go to complete the journey. Normally in a HEV, the ICE
would be required to complete the journey. However a second even lower SOC
limit is allowed for the vehicle to complete the journey.

2.3 Control for HEV ZEV Zones

There are plans emerging for ZEV zones within city centres which present new problems particularly for HEVs. It may be that the HEV cannot complete its traversal of a ZEV zone if it does not have enough battery charge. Prior knowledge of when a HEV is about to enter a ZEV zone can mean that the ICE is used to change the battery to upper SOC immediately before entering the ZEV zone.

2.4 Optimised Use of EV/HEV Ancillaries

If an EV or HEV is nearing the lower limit of battery SOC but not quite nearing the end of its journey, one possible strategy is the limitation of vehicle acceleration so that an estimated journey can be completed without running out of charge. Another is reducing the loading required by ancillaries such as power steering, power braking, HVAC etc. A further one is for the control system not to act as a controller but advising the driver that he can reduce load on the battery by switching off radio, heating, demister and air-conditioning. With all of these the challenge is to do them in such a way so that it is acceptable to the customer.

2.5 HEV/EV Energy Consumption Rate Based Control

The energy consumption parameter is a function of the conversion efficiency of the energy stored in the battery to electric current, the efficiency of the power electronics, the efficiency of the driving motor(s), the efficiency of the torque transmission system and a multitude of efficiencies which are variables and due to ancillaries (e.g. HVAC etc). From knowledge of average consumption and the battery SOC and how much fuel (in case of HEV), the driving range of the vehicle can be calculated. Therefore if the distance to the next destination of the vehicle is known and the overall consumption rate is given then the required energy to be stored in the vehicle can be estimated. In the ideal situation the destination and the actual route would be known, therefore the distance can be accurately determined. In practice the average consumption rate cannot be accurately known from the beginning of the journey since it is the result of the integration of the differentiated instantaneous consumption rate per unit distance throughout the entire journey, and subjected to a multitude of causes of variability (e.g. traffic conditions, driving style etc). It is therefore desirable to estimate the energy requirements of a journey and it is assumed that the distance of this journey is given or can be measured, a way should be found to minimise the variability in the estimation of the average energy consumption rate.

An approach to journey optimised control based on a thermostatically controlled series HEV is described here to demonstrate the use of consumption rate. If the journey destination is known, the overall consumption rate of a hybrid vehicle is unlikely to be correctly predicted from the beginning of the journey and therefore it must be measured during the operation of the vehicle. The average consumption rate can be calculated over predetermined intervals of the journey (e.g. 100m). The average value is considered to continue to apply in the next interval of the journey. Once the interval is traversed the actual consumption rate can be measured and a new assumed consumption rate can be calculated. This procedure is repeated until the vehicle reaches its destination. This strategy can be enhanced by the use of GPS location information for determination of the vehicle location at the end of each interval along the journey route.

3   Investigation into Journey Parameter Prediction

The previous sections have motivated the need for journey energy parameter prediction in the powertrain controller of a HEV. This section outlines an investigation into the feasibility of journey parameter prediction using data collected from a single vehicle over a period of one month. The first part of the investigation considers that the only available indicator variable with which to make the prediction is departure time, since this type of information could easily be made available in a modern passenger vehicle. The second part of the investigation explores the benefits to prediction accuracy of the inclusion of place of departure information that could be supplied by a GPS (Global Positioning System) receiver. Five parameters are required for the investigation and can be derived from a speed / time profile. The parameters of interest as mentioned in the previous section are energy requirement, energy recoverable, stationary period, distance, duration, GPS vehicle location information.

3.1   Data Collection

Data was collected using a GPS receiver connected to a laptop PC. A vehicle data logging programme was set up for the collection of vehicle usage data from different vehicles. Ten vehicles were chosen based on the age and sex statistics of the main vehicle user, such that the vehicles formed a simple representation of the UK driving population. However, the data from only one of these vehicles is presented in this paper since it was the only one with the brake light instrumented so that energy recoverable can be derived and it was the subject with the most predictable vehicle usage characteristics out of the vehicles recorded and therefore showed the potential of optimal HEV/EV control. Further information on the design of the vehicle data logging programme and investigations into data from the other vehicles can be found in previous work by the author [9, 10].

3.2 Assessment of Journey Energy and Recoverable Energy

Energy requirement is the most important parameter for journey parameter optimised HEV control. If an example vehicle has a battery with a capacity of 5.6 KW/h, this in theory gives a total energy available of 20.16 Mjoules. However due to high battery charge and discharge resistance, and the possibility of irreparable battery damage, the battery SOC never reaches either 0% or 100%. Typical lower and upper switching SOC values are 10% and 90% respectively [5]. Therefore assuming that this leaves the area of battery operation in a region of linearity, the amount of energy for traversal in ZEV mode only is 16.128 Mjoules.

The speed / time profile taken from the data logger were summarised into the two required journey parameters. Both energy requirement and energy recoverable can be derived from the road load equation which includes the four components acceleration force, rolling resistance, aerodynamic drag and resistance due to gradients and calculating the instantaneous power (P), which is the product of road load and velocity:

$$P = (M * a * V) + (C_r * M * V) + (0.5 * pC_d * A * V^3) + (M * sin(\alpha) * V) \qquad (3)$$

Only a speed / time profile was available and therefore resistance due to gradients was assumed to be zero. This was deemed reasonable since the data was collected in an area with little change in gradients. Additionally, if the vehicle which the data is collected from is driven over a repeatable route on its regularly occurring journeys, then the error due to omission of the gradient components should be reasonably consistent. Therefore the appearance of predictable patterns should not be affected significantly. Therefore the road load equation used for the derivation of the power / time profile only consists of the first three terms in (3).

The energy components can be obtained from the integral of the power / time profile. The energy has two main components; energy requirement and energy loss. The area above the axis (at instantaneous power = 0) is the energy requirement of the journey. The area below the axis is the energy loss during vehicle deceleration. To obtain values for energy recoverable for each journey, the area of the energy loss component coinciding with the brake light signal at high is used. Figure 2 compares the power / time profile of a small journey segment with the light flag signal. It can be seen that the brake light signal corresponds with parts of the power / time profile below the Instantaneous Power = 0 axis, which are the energy recoverable. The areas shaded in grey are the energy recoverable derived from the negative parts of the power / time profile coinciding with the brake light signal at high. The areas shaded black are the energy requirement derived from the positive parts of the power / time profile. This shows that the energy requirement and energy recoverable may be quantifiable from the speed time profile, therefore easily obtained

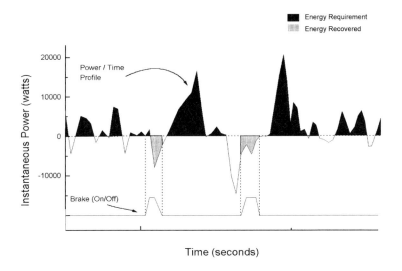

Fig. 2.   The power / time profile compared with the brake light signal for a small segment of a single journey

3.3   Estimation of Distance, Duration and Energy Requirement from Time of Day

Initial data analysis involved visualising each of the parameters in turn with respect to departure time. From Figure 5, it can be seen in the 'week days' plot that there is one significant peak representing journeys occurring around 7 a.m. with duration and distance of around 1200 seconds and 13km. Separation of the data into weekend and week day subsets showed that there are no discernible patterns for the weekend journeys. Therefore, for this vehicle user, weekend journeys appear unpredictable. However week day journeys do show a very definite element of predictability. Data visualisation for the other journey parameters gave similar results. As a result an investigation into the prediction of the journey parameters during the week days only is considered further. This initial visual analysis shows that the journey parameters distance, duration and therefore energy requirement could quite possibly be predicted for this vehicle user's week day morning journeys by the use of some simple rules:

If (departure time is around 450 minutes) then (energy requirement will be around 5 Mjoules) AND (distance will be around 13km) AND (duration will be around 1200 seconds)

This heuristic approach to rule development does still have the unsolved problem if a similar technique were to be implemented in a HEV/EV of how to determine which journeys are predictable and which are not. Inclusion of journey location information and analysis of energy consumption rate at regular intervals may provide a way of dealing with this. If a journey prediction method of a similar kind were to be implemented in a hybrid electric power train controller, an automatic method of rule generation would be needed. So far there has been no analysis of the prediction of the parameters energy recoverable and stationary period.

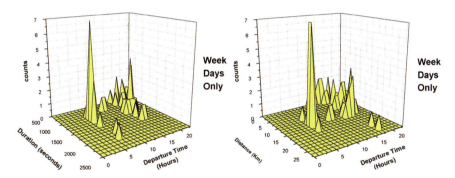

Fig. 3.   The relationship between distance, duration and departure time

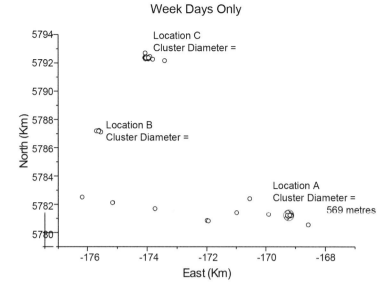

Fig. 4.   Locations visited expressed as displacement from the origin of the earth's co-ordinates

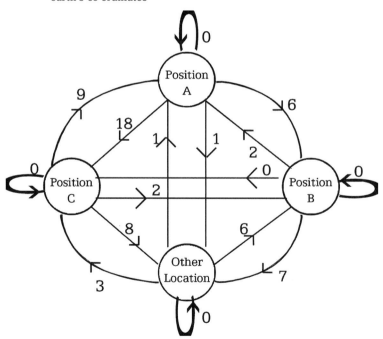

Fig. 5.   Nodal analysis of locations regularly visited for week days (* numbers at each arc represent the number of journey occurrences)

3.4   Location Information Based Journey Parameter Estimation

Figure 4 shows the location of journey departure plotted for each of the week day journeys for the subject. It can be seen that there are three regularly visited locations present, each described by a cluster of data. Each cluster had different radii which was due to noise inherent in each of the locations such as that caused by variability in vehicle parking location and noise that was inherent in the GPS signal. The origin and destination clusters were examined further in terms of the number of occurrences and energy requirements. The number of occurrences of the journeys between these locations is shown in Figure 5. Table 1 shows the summary of the journeys between the regularly visited locations along with the mean energy requirements of the journey. It can be seen that three of the regular journeys have variation in energy requirements that are less than ten percent, but by referring to Figure 5, it can be seen that these are the journeys that occur the least. The most frequently occurring journeys are between locations A and C and the energy requirements can vary by up to thirty percent.

Journey	Mean Energy (joules)	Std. Deviation (joules)	Std. Dev. as a%
A to C	5,178,192.70	947,986.64	18.3%
C to A	5,208,464.86	1,528,553.76	29.3%
B to C	-	-	-
C to B	2,389,878.75	78,878.00	3.3%
A to B	2,178,160.93	155,503.31	7.1%
B to A	2,522,667.33	210,262.89	8.3%

Tab. 1.   Journeys and energy requirements

This can lead to the following rule for energy requirement estimation: If (week day) & (departure is location B) & (departure time is around 450 minutes) then (energy requirement is around 5 Mj). However, there is a potentially large variation in energy requirement that would have to be dealt with for optimal control. This could be caused by varying traffic conditions and route.

4   Conclusion

A number of control system scenarios and strategies have been presented for the optimal control of HEVs and EVs, some of which require journey location and route information for their successful operation. Preliminary results have been presented on whether there is the possibility to predict journey parameters such as distance, duration, energy requirements, stationary period and energy recoverable at the start of a journey. For regularly occurring journeys distance and duration and energy requirement seem reasonably repeatable. The journey parameters energy recoverable and stationary period require further study. It is clear that there is enough variability in the energy requirements of commonly occurring journeys such that route and traffic information will be required in order to have an algorithm to deal with these uncertainties.

Further work should address a number of issues. It is clear that route information is necessary to be able to estimate the energy requirements of a journey more accurately and also help in the continual re-evaluation this estimation throughout the journey. Journey energy requirements could also be useful information to utilities companies who need to supply electricity from the grid for the charging of plug-in HEVs and EVs.

References

[1] Andres, D.J., Guziec P.R., Weinstock R.A.; "University of Illinois Hybrid Electric Vehicle Philosophy and Architecture", SAE SP - 980, Burke A.F. and Smith G.E. (1981); "Impacts of Use Pattern on the Design of Electric and Hybrid Vehicles", SAE Paper 810265, 1993.

[2] Byl, M., Cassanego P., Eng G., Herndon T., Kruetzfeld K., McKee A., Reimers G., Frank A.; "Hybrid Electric Vehicle Development at the University of California, Davis : The Design of Aftershock", SAE 1995 SP – 1103, 1995.

[3] Farrall, S.D.; 'A study in the use of fuzzy logic in the management of an automotive heat engine /electric hybrid vehicle powertrain', PhD thesis, Dept. of Engineering, University of Warwick, UK, 1992.

[4] Farrall, S.D. and Jones, R.P.; 'Energy management in an automotive electric/heat engine hybrid powertrain using fuzzy decision making', Proceedings of 1993 IEEE International Symposium on Intelligent Control, pp 463-468, Chicago, Aug. 1993.

[5] Farrell, J.A. and Barth, M.J.; 'Hybrid electric vehicle energy management strategies', Final report prepared for ISE Research, 4909 Murphy Canyon Road, Suite 330, San Diego, CA 92123, USA, 1996.

[6] Kalberlah, A; 'Electric hybrid drive systems for passenger cars and taxis', SAE paper 910247, 1991.

[7] Malasiotis, E. N.; "Prediction of the Journey Energy Requirements for the Optimised Control of a Hybrid Electric Vehicle", Master of Science Thesis, University of Warwick, Coventry CV4 7AL, UK, 1998.

[8] Quigley, C.P.; "Prediction of Journey Parameters for the Intelligent Control of a Hybrid Electric Vehicle", M.Sc Thesis, Dept. of Engineering, University of Warwick, Coventry, United Kingdom, 1997.

[9]   Quigley, C.P., Ball, R.J., Jones, R.P.; "Fuzzy Modelling Approach to the Prediction of Journey Parameters for Hybrid Electric Control", Proc. of the Institution of Mechanical Engineers, Vol.214, Part D., 2000.

[10] Wyczalek, F.A. and Wang, T.C.; 'Regenerative braking concepts for electric vehicles - a primer', SAE paper 920648, 1992.

Terms Used

E_A is total energy available for traversal

E_B is energy in battery at start of journey

E_R is energy that can be recovered by regenerative braking

E_F is energy that can be supplied by the ICE which is dictated by the amount of fuel in the fuel tank at the start of journey

ε_B is efficiency of the conversion of battery energy to mechanical torque through the electric motor at the transmission

ε_R is efficiency of recovering lost energy and replacing it in the battery

ε_{ICE} is efficiency of the ICE

M is vehicle weight (Kg)

a is vehicle acceleration (ms-2)

V is vehicle speed (ms-1)

C_r is coefficient of rolling resistance

p is air density (Kgm-3)

C_d is vehicle drag coefficient

A is vehicle cross-sectional area (m2)

a is incline of a road gradient (radians)

Chris Quigley, Richard McLaughlin
Warwick Control
8 Ladbroke Park, Millers Road
CV34 5AN Warwick
United Kingdom
chris@warwickcontrol.com
richard@warwickcontrol.com

Keywords: Electric Vehicles (EV), EV Journey Prediction, EV Control Systems, Hybrid Electric Vehicles (HEV), HEV Control Systems, Journey Energy Requirements, Zero Emission Vehicles

EcoGem - Cooperative Advanced Driver Assistance System for Green Cars

M. Masikos, K. Demestichas, E. Adamopoulou, NTU Athens
F. Cappadona, Pininfarina S.P.A.
S. Dreher, NAVTEQ B.V.

Abstract

This paper advocates that the success and user acceptability of Fully Electric Vehicles (FEVs) will predominantly depend on their electrical energy consumption rate and the corresponding degree of autonomy that they can offer. FEVs must provide their drivers with the highest possible autonomy as well as with a high degree of reliability and robustness in terms of energy performance. The FEV driver must know at all times, with a high degree of assurance, if a destination is reachable with the remaining battery energy, how much energy will be left after his journey, how to efficiently reduce the energy required to reach his destination, as well as when and where it is better to recharge his vehicle. This paper argues that appropriate innovative ICT solutions must be pursued and adopted, in order to assist the driver in dealing with such energy-related issues, and to strengthen FEVs' autonomy and reliability. Such an ICT solution tailored for FEVs –entitled EcoGem– is described in this paper.

1   Introduction

Over the last ten years the progressive integration of Advanced Driver Assistance Systems (ADAS) first in high-end cars and lately also in intermediate vehicles has contributed to the improvement of the safety and comfort of motorists. A recently released report from the U.S. Department of Transportation's National Highway Traffic Safety Administration (NHTSA) revealed a decline of 9.7% of road deaths from 2008 to 2009 [1], despite increasing overall vehicle travel, which they claim coincides with the increase of the availability of ADAS systems.

This paper presents a new ADAS system entitled EcoGem, which is a tailored-for-Fully-Electric-Vehicles (FEVs) ADAS, equipped with suitable monitoring, learning, reasoning and management capabilities that will help increase a

FEV's autonomy (distance that can be travelled before battery depletion) and overall electrical energy efficiency.

To this end, EcoGem's approach is based on the following two primary goals:

▶ To render the FEV capable of reaching the desired destination(s) through the most energy efficient route(s) possible;

▶ To render the FEV fully aware of the surrounding recharging points/ stations while travelling.

The first goal will allow for a reduction of the amount of energy that is spent to reach the driver's desired destinations, leading to an increase in the vehicle's autonomy. Actual reduction rates can vary according to the actual routing options at hand. Nonetheless, a significant reduction can be achieved in cases in which a heavily congested route is avoided, and a noteworthy reduction can also be accomplished when a route comprising steep roads (i.e., high slopes) is bypassed. The second goal will make the everyday use of an electric car more robust, enabling efficient scheduling of battery recharges and preventing battery depletion while on the move. This is greatly important for a large percentage of plug-in electric vehicles, especially those whose domestic recharging is not an option for their owners or is found insufficient.

The remainder of this paper is structured as follows: Section 2 provides an analysis of ADAS systems and categorizes the ADASs developed up to today. A more detailed presentation of the state-of-the-art ADASs is given in Section 3. The proposed EcoGem ADAS is described in detail in Section 4. More specifically, this chapter reports all the functionalities that are provided by the EcoGem ADAS in order to fulfil its purpose. The paper is finalised with a conclusion chapter.

2   Advanced Driver Assistance Systems (ADAS)

ADASs are in-vehicles systems that allow the driver to reach his destination in a less stressful, safer, more comfortable and efficient way. Most of the Driver Assistance Systems currently on the market are mainly comfort applications, aiming at reducing driver workload and allowing the driver to focus more on secondary tasks, or safety applications, aimed at supporting the driver in dangerous situations by preventing or minimizing collisions. ADAS systems can be divided in four categories:

▶ Active Safety (Adaptive Front Lightning, AFL; Traffic Sign Recognition, TSR; Forward Collision Warning,FCW; and Collision Mitigation)

▶ Driver Support (Curve Speed Warning,CSW; Lane Departure Warning,LDW)

▶ Driver Information (Speed information and Speed Limit Advisor, Black Spot Warning available in Japan)

▶ Energy Efficiency (Adaptive Cruise Control, ACC; Powertrain Efficiency, Fuel Efficiency Advisor, FEA; Eco Routing and Eco Driving)

The safety related ADAS systems can also be classified into categories according to the time before a potential impact, as shown on Figure 1. These categories are navigation period, manoeuvring period, stabilisation period, period before crash, and period after crash.

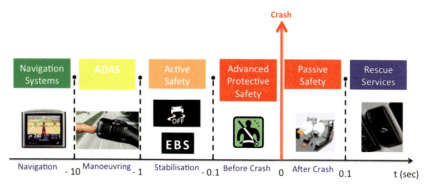

Fig. 1.   Categorisation of safety driver assistance systems according to the time relative to a crash situation

3   State-of-the-Art

Since 2000, ADAS systems have been the main focus of EU-funded research, mainly in the framework of the eSafety initiative [2] and the EU-funded SafeSpot project [3].

The eSafety initiative was officially launched by the European Commission in April 2002. The scope of this initiative is to bring together the European Commission, Member States, road and safety authorities, automotive industry, telecommunications industry, service providers, user organisations, insurance industry, technology providers, research organisations and road operators to accelerate the development, deployment and use of eSafety systems. eSafety systems are intelligent vehicle safety systems that use information and communication technologies in intelligent solutions, in order to increase road safety and reduce the number of accidents on Europe's roads.

The aim of the EU-funded SafeSpot project was to develop a Safety Margin Assistant that detects in advance potentially dangerous situations and that extends in space and time drivers´ awareness of the surrounding environment and provides warnings in critical or dangerous situations on an on-board HMI. SafeSpot has developed a set of technologies like data fusion to combine information from different vehicle sensors and roadside sensors and the Local Dynamic Map, a dynamic data store for safety critical location-based information, in order to enable intelligent cooperative systems capable of preventing road accidents. Several prototype vehicle and infrastructure-based applications have been developed based on the SafeSpot core technologies to demonstrate the potential of the SafeSpot system. These applications constitute realistic examples of what could be the basic set of safety-related services available for connected vehicles in the near future and include Road Intersection Safety, Lane Change Manoeuvre, Safe Overtaking, Head On Collision Warning, Rear End Collision, Speed Limitation and Safety Distance, Frontal Collision Warning, Road Condition Status – Slippery Road, Curve Warning and Vulnerable Road User Detection and Accident Avoidance.

The market adoption of these systems is however still very low due to the use of expensive sensors like radars. However, today several driver assistance solutions can be provided based on simple cameras with sophisticated image processing techniques. Mobileye [4] is for example developing a vision-based ADAS targeted at collision avoidance and mitigation. Their image processing techniques can support applications like Vehicle detection, Forward Collision Warning, Lane Departure Warning, Pedestrian detection, Traffic Sign Recognition and Intelligent Headlight Control. Mobileye is currently also providing a standalone portable device with built-in camera for autonomous recognition and warning applications.

3.1 Eco-Driving Enabled ADAS

Eco-driving assistance systems are applications that encourage economical driving behaviours and support the driver in optimizing his driving style to achieve fuel economy and consequently emission reduction. They constitute an essential feature for hybrid vehicles and have been introduced by several manufacturers for their hybrid vehicle range. For example, the Honda Insight Eco Assist is a sophisticated feedback system that teaches the driver to drive more efficiently; Ford Fusion and Mercury Milan SmartGauge display eco-driving related information along with fuel and battery power levels; and

Fiat eco:Drive provides the ability to record the journey on a USB stick while driving and allows checking journey data after the trip on a computer with the possibility to track the progress under the form of a challenge with an eco index.

In this context, navigation system manufacturers have recently introduced the first driver assistance systems with support for fuel economy. For example, Garmin provides the ecoRoute software for its nüvi devices, which allows the driver to calculate the most fuel-efficient route, based on road speed information and vehicle acceleration and not on map data, while TomTom provides a device called ecoPLUS that measures and benchmarks vehicle fuel efficiency and carbon footprint, and Crambo labs provide the Econav software, which gives real time indications about efficient gear and optimal speed to achieve up to 30% fuel and CO2 savings.

3.2 Map-Enabled ADAS

Sensors commonly used in ADAS applications help provide relevant contextual information to a vehicle, including the road conditions in front of the vehicle. When compared to the depth of road information available in a map, however, the range of these sensor inputs is confined to a relatively close proximity around the vehicle's path. In many cases, map attributes, like geometry, curvature, height/slope, speed limit or lane information, provide enabling content or can enhance the performance of Driver Assistance Systems. Map attributes can be thought as a type of sensor that dramatically extends existing sensors' range. Traffic lights and stop signs can be additionally useful for eco-driving or eco-routing applications. Applications where the Map attributes are necessary include Speed Limit Warning, Hazard or Black Spot Warning, Curve Speed Warning, Dynamic Pass Prediction, Eco-driving based on road geometry, Adaptive Front lights, Predictive Cruise Control and Hybrid Powertrain Management. Commercial products that exploit Map attributes are NAVTEQ's Green Streets map product that provides green driving and green routing applications, along with Traffic patterns and Real-Time Traffic information, the ACC system of BMW launched in 2005, and the RunSmart Predictive Cruise Control (PCC) system that Daimler launched for its Freightliner Trucks in 2009. The RunSmart system evaluates the upcoming road profile one mile ahead on the route and anticipates road grades by using GPS and 3D digital map technology. The system adjusts the actual speed of the truck for maximum fuel efficiency based on the terrain while staying within 6% of the set speed. Another solution of NAVTEQ that is independent of the vehicle's navigation system and dedicated to map-based ADAS is the Map and Positioning Engine (MPE) [5], which consists of an embedded ADAS map and the Electronic

Horizon used for applications ranging from driver assistance to active safety and enhanced fuel economy.

4   EcoGem ADAS

All of the ADAS systems described previously were developed primarily to enhance driver safety and comfort, and some of them secondarily to help reduce fuel consumption. The EcoGem project goes one step further by proposing an ADAS especially designed for Fully Electric Vehicles (FEVs) [6]. FEVs present some special characteristics due to their electrical nature. The fact that their energy storage capabilities are limited, and that their recharging time is not as low as the time it takes to fill a fuel tank, has a major effect on their autonomy. The EcoGem ADAS is especially designed in order to provide some extra functionality that ensures comfortable and relaxed driving. To be more precise, the EcoGem ADAS attempts to eliminate driver anxiety regarding the distance to charging station, the next time of charging, and the destination reachability, and aims at significantly extending FEVs' range.

The EcoGem ADAS renders the FEV capable of reaching the desired destination(s) through the most energy efficient route(s) possible via either autonomous optimised route planning or cooperative optimised route planning. In both planning schemes, the most energy efficient route calculation is achieved by applying machine learning algorithms on past tracked data records. These data records include energy consumption information per entire route and per route segment, time-related information like time-zone and month, vehicle-specific information like battery info and consumption rate, map-related information like road segment inclination and length, and weather-related information like humidity. In case of autonomous routing, the EcoGem ADAS exploits the vehicle's own tracked records, while in case of cooperative routing it exploits vehicle-to-vehicle (V2V) interactions to share its route-selection experiences with other EcoGem FEVs. It is obvious that in the cooperative case the route planning is more efficient as it is based on a wider set of past experiences. Moreover, this community-based approach may exploit also vehicle-to-infrastructure (V2I) communication in order to allow sharing of the gathered experiences with the management infrastructure, enabling thus efficient fleet management, which is particularly important for electric vehicle fleets (such as electric bus fleets, or electric van fleets, etc.). Centralised fleet management can complement the distributed mode of route planning, in achieving the most efficient routing decisions possible.

The EcoGem ADAS is also responsible for continuous awareness of recharging points and optimised recharging planning. Based on the current battery levels, energy consumption rate and contextual information (desired destination, present location, daytime, traffic, user agenda, etc.), the EcoGem ADAS can prompt the driver, whenever necessary, to select a recharging option (normal or fast recharging, or battery replacement) and to book the most convenient recharging point. Booking in advance allows for exclusive access to the recharging point at the time of arrival. The ADAS must ensure that the recharging point is reached on time, by informing the driver about the optimal route to the recharging station, and must also minimise the detour caused to the driver. Recharging management is a very critical issue considering the sensitivity and vulnerability of rechargeable electrical batteries. Timing of charging, state-of-charge (SOC) level at the moment of charging, SOC target at the end of charging process, frequency of charging, they all affect the lifetime and usable energy level of the battery. Thus, EcoGem provides an important tool for both the vehicle and battery manufacturer to predict much more precisely the SOC levels at distances ahead. The data regarding the availability of the nearest charging stations, as well as the traffic and road conditions data in alternative paths, are provided by the EcoGem system. This gives the ADAS the input required to choose the best alternative station, taking into account the minimum charging level required, the actual condition of the battery cells and the amount of time available for charging. Thus, alternatives among fast charging (if available), standard charging (the percentage of the target charge level is a decision to be taken by the system), or other method (e.g., battery replacement), can be chosen based on reliable and actual data.

The functionalities encompassed by the EcoGem system, in order to provide the previously mentioned ADAS services, are depicted at the EcoGem concept architecture figure (Fig. 2) and can be summarised as follows:

> ▶ In-vehicle functionality: (i) on-going on-board collection of measurements (i.e., travelled routes, time spent on each route, battery consumption, etc.); (ii) on-going collection of measurements and knowledge extracted by cooperating EcoGem vehicles or received from the central platform; (iii) secure storage of the gathered, multi-source measurements and information; (iv) learning and reasoning on the available information for automatic classification of potential routes according to their degree of congestion (traffic estimation), which allows for optimal route planning; (v) extraction of policies, i.e. traffic-related conclusions and decisions according to the current context, based on past experience; (vi) sharing of local information with the platform and other EcoGem vehicles; (vii) energy-driven optimised route planning based on road characteristics from the ADAS map database (e.g., road slopes) and the estimated degrees of congestion; (viii) continuous awareness of

surrounding recharging points (estimated distances, availabilities, etc.); (ix) design of the best recharging strategy, based on available battery levels, current position, desired destination, recharging point availability, traffic, daytime, user agenda, etc.

▶ Central platform functionality targeted for: (i) collection of measurements and knowledge from the EcoGem-enabled vehicles; (ii) integration and joint exploitation of additional traffic data originating from other, alternative traffic monitoring systems; (iii) large-scale traffic prediction, through processing of gathered multi-source information; (iv) optimal traffic distribution (traffic optimisation), based on the traffic estimations; (v) optimal traffic control/management actions (instructions and suggestions) and enhanced traffic information services; (vi) online monitoring and management of recharging points; (vii) provision of access to the rich information of the platform (service-enabling functionality) to external entities, such as end-users and electrical recharging point providers.

▶ V2V and V2I interactions for conveying: (i) recent measurements (recently acquired traffic data that most probably are still valid); (ii) historical measurements (that mainly serve as training data for the machine-learning engines of the EcoGem vehicles); (iii) knowledge (route classifications, route selections, derived policies) extracted through machine-learning based processing. Additional V2I/I2V interactions are responsible for enabling awareness of recharging options and booking of recharging points, while I2V one-way interactions may provide traffic control instructions and suggestions.

Based on the functionality described above, the EcoGem ADAS calculates several routes to the desired destination based on different parameters. For example, the EcoGem ADAS may calculate either the route with the minimum travel time, or the most energy efficient route. These alternatives are presented to the driver via a user-friendly interface and he / she is responsible for the final choice. Besides this, the driver's choice may cause the generation of an alternative recharging strategy. Furthermore, the EcoGem ADAS may recommend modifications in the operation of the electrified auxiliaries, in order to reduce vehicle's energy consumption and enable it to reach the desired destination.

The selections that are made by the driver each time may be saved in order to constitute his/her profile characteristics. A selected route may be saved to the EcoGem ADAS in order to allow the driver to load it at a later time and avoid new calculations. Of course, in such a case the loaded route may not be up-to-date regarding current conditions. This functionality imposes the necessity for support of multiple user profiles as several drivers may use the same vehicle. Each driver usually has his/her own set of preferences and saved routes. Thus, the EcoGem ADAS shall recognise each driver in order to load the corresponding preferences and routes, and make the trip more comfortable.

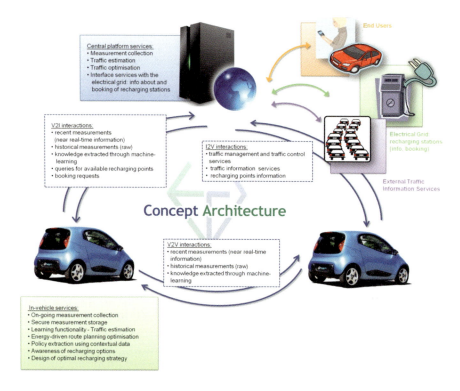

Fig. 2.   EcoGem Concept Architecture

The user profiling feature of the EcoGem ADAS also enforces the development of security and privacy mechanisms. User identity authentication is necessary in order to activate the EcoGem ADAS and load user preferences and saved routes. This authentication can be accomplished through various methods, including passwords, tokens, smart cards and biometrics. Secure data storage is also ensured; the ADAS databases are secured from any manipulation or modification caused by users or applications that are not authorized to do so. The ADAS user privacy is also preserved, as the user identity is neither recorded in the ADAS databases nor associated to any recorded data.

5   Conclusion

The present paper constitutes an important report describing the evolution and development of ADAS systems. It presents the progressions made in this area and cites the most representative ADASs that have been developed up to now. Initially, the primary functionalities offered by ADASs aimed at making the driving experience more comfortable and relaxed. Subsequently, the rapid increase of the vehicles' volume and of the number of car accidents significantly affected the goals of ADASs. New systems were developed aiming initially at accidents prevention, then at injuries mitigation and finally at rescue time minimization. These safety features of the ADASs have forced more and more vehicle manufacturers to adopt them and offer them as standard equipment.

The EcoGem project, after taking into account this evolution regarding ADASs, as well as the evolutions taking place in ICT technologies, proposed the development of an ADAS specialised for mitigating the range-anxiety of FEVs. Particularly, the EcoGem project attempts to extend FEVs' autonomy and to render their usage more robust via presenting an ADAS equipped with enhanced energy and traffic monitoring and management functionalities. These functionalities are further reinforced by EcoGem's cooperative operational mode, either in a V2V or in a V2I manner.

Future work includes testing of the EcoGem ADAS both in a simulated environment (using a micro-simulation method) as well as in trial sites using real vehicles. Vehicles that will be used include Pininfarina's Nido, an urban 2-seater FEV, and TEMSA's electric bus. Trials will help systematically quantify the benefits stemming from the use of the EcoGem ADAS, as well as make any improvements necessary for promoting user adoption.

References

[1]   Goodwin A., et. al., 'Countermeasures that Work', the U.S. Department of Transportation-National Highway Traffic Safety Administration, Washington, 2010.

[2]   eSafety website URL: http://www.icarsupport.org/

[3]   Safespot website URL: http://www.safespot-eu.org/

[4]   Mobileye website URL: http://www.mobileye.com/

[5]   NAVTEQ website URL: http://www.navteq.com/

[6]   The EcoGem Project website URL: http://www.ecogem.eu/

Konstantinos Demestichas, Evgenia Adamopoulou, Michalis Masikos
National Technical University of Athens
Heroon Polytechneiou, 9
15773 Zografou, Attiki
Greece
cdemest@cn.ntua.gr
eadam@cn.ntua.gr
mmasik@telecom.ntua.gr

Filippo Cappadona
PININFARINA S.P.A.
Via Nazionale, 30
10020 Cambiano, Torino
Italy
f.cappadona@pininfarina.it

Stephane Dreher
NAVTEQ B.V.
Minervastraat, 6
1930 Zaventem
Belgium
stephane.dreher@navteq.com

Keywords: advance driver assistance system, ADAS, full electric vehicle, FEV, FEV optimised routing, FEV recharging strategy, extending FEV autonomy

Human-Centered Challenges and Contribution for the Implementation of Automated Driving

A. P. van den Beukel, M.C. van der Voort, University of Twente

Abstract

Automated driving is expected to increase safety and efficiency of road transport. With regard to the implementation of automated driving, we observed that those aspects which need to be further developed especially relate to human capabilities. Based on this observation and the understanding that automation will most likely be applied in terms of partially automated driving, we distinguished two major challenges for the implementation of partially automated driving: (1) Defining appropriate levels of automation, and; (2) Developing appropriate transitions between manual control and automation. The Assisted Driver Model has provided a framework for the first challenge, because this model recommends levels of automation dependent on traffic situations. To conclude, this research also provided brief directions on the second challenge, i.e. solutions how to accommodate drivers with partial automation.

1   Introduction

Automated vehicles are, compared to human drivers, superior with respect to precision of operation and ability to operate under severe circumstances. Automated cars are therefore expected to cause less accidents and reduce congestion [1]. These advantages have the potential to help achieving goals for safer and more efficient road transport as set by the European Union [2]. However, autonomous driving involves more than automating the operational task alone. Generally, its realization can be divided in the areas: Navigation; Sensor technology (observing and understanding the vehicle's direct environment); Decision making (planning the vehicle's direct path of motion appropriate for the immediate traffic situation), and: Actuation (i.e. operating the vehicle). The areas Navigation and Actuation are very well developed. Most effort is currently addressed towards the sensor technology. The least developed area is: Decision making. Remarkably, this area is probably also the most difficult to solve. Due to the diversity in traffic situations and variety in traffic participants' behavior, it is very difficult to interpret and precisely predict oncoming changes in traffic situations. Interestingly, it is especially at these interpretation and decision making tasks that humans are generally good at in comparison to machine operation [3]. Recent demonstrations with automated vehicles on public roads – i.e.: The Stadtpilot project in Braunschweig, the Vislab Intercontinental Autonomous Challenge and Google's 'Robotic Cars' project– illustrate this state of art: Although the projects show far reached technical capabilities for automated driving, each of them also reports the necessity for human intervention in complex driving situations, like merging lanes or crossing an intersection.

The above explained state of art for automated vehicles illustrates that the development is most of the times based on what is technologically possible, not necessarily on what drivers are in need for [4]. Therefore, this research is intended to contribute to the development of autonomous driving by considering a human-centered approach. To do this, the next chapter will first explain our estimation upon the scale of implementation for autonomous driving, i.e.: the implementation of partially automated driving instead of complete automation. Based on this view, the chapter will also explain why two major challenges for the realization of automated driving relate to human aspects, i.e. (1) Defining the appropriate levels of automation, and (2) Developing appropriate transitions between manually and automated driving (vice versa). As an attempt to define appropriate levels of automation, chapter 3 introduces an Assisted Driver Model, which recommends driving support dependent on driving situations. The last chapter will comment on the aspects involved in designing the transitions between manually and automated driving.

2 Motivation for and Challenges of Partially Automated Driving

Current applications of (completely) autonomous driving are practiced within closed environments and with the support of dedicated infrastructure. Examples are driverless container terminals in harbours or driverless taxi's at airfields. For the future, people might envision autonomously driving vehicles which merely replace current passenger vehicles and make use of existing infrastructure. Following the autonomous vehicles' state of art from the introduction, the next section will explain why partially automation is a more realistic view for large scale implementation of automated driving than completely automation of the driving task. After that, the second section continues our considerations how human aspects relate to the implementation of autonomous driving and their subsequent challenges.

2.1 Motivation for Development of Partially Automated Driving

The first reason why the implementation of partially automated driving is regarded more realistic than complete automation relates to the fact that humans are more capable of dealing with the diversity in traffic situations, driving circumstances and road users. Secondly, due to technical constraints there will always exist system boundaries. Therefore, the system design needs to account for exceeding these boundaries, i.e. takeover by human operation. A third reason relates to liability: Drivers are personally liable for safe driving. In case something goes wrong, drivers need therefore be able to take over full control at any moment. On top, complete automation does not seem desirable, as it diminishes one of the automobile's remarkable attributes: i.e. the fun of driving and mastering a vehicle. A realistic view for applying autonomous driving is therefore: partially automated driving. Within this view, we acknowledge two general possibilities for partially automated driving: (1) The automation of a specific driving task, e.g. the automation of way finding with the aid of a navigation system, and (2) Applying automation to specific traffic situations, e.g. automated parallel parking. Both possibilities are visualized in Figure 1. The main differences are the involved time span versus level of automation. For traffic situations the involved level of automation might be high, but for a limited period of time. For driving tasks, the level of automation might be low, but involve a longer time span. The machine does not acquire continuously full control and the human driver will need to be part of the control-loop on a frequent basis. This view on the realization of automated driving is in line with a previous assessment of the implementation of automated and semi-automated transport systems [5].

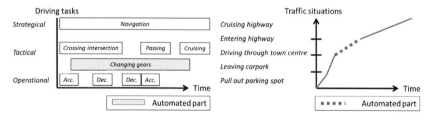

Fig. 1.   Examples of partially automation applied to respectively driving tasks (left) and traffic situations (right)

2.2 Challenges for Development of Partially Automated Driving

The previously unveiled view of partially automated driving and the importance to consider human capabilities when developing solutions for the implementation of automation, lead us to assuming two major human-centered challenges: This is on the one hand defining the appropriate level of automation and on the other hand developing appropriate transitions to change between manually and automated driving (and vice versa).

A consequence of applying automated driving to specific driving situations or tasks is that transitions to and from these modes need to be accommodated. That means that appropriate solutions for giving and retrieving control need to be developed. Human Factors concerns, related to partial automation, underline the importance of appropriate transitions. The concerns are especially related to out-of-the-loop (OOTL) performance problems [6]. These problems basically mean that a user (the operator) is placed remote from the control loop during a situation of automated driving. As a consequence, the operator's awareness of the situation or system's status may be reduced. This causes problems for transitions to and from manual operation (especially when system errors, malfunction or break-downs occur), resulting in slower reaction times, misunderstanding what corrective actions need be taken and manual skill decay [7]. This underlines the importance of the second challenge, i.e.: developing appropriate transitions between manual and automated driving (vice versa).

3 Defining Appropriate Levels of Support for Partially Automated Driving

As an attempt to help reducing the first challenge, i.e.: defining the appropriate level of automation, this chapter answers the following questions: What driving situations can be distinguished?; What levels of automation should be distinguished?, and: What automation level is recommended for which driving situation?

3.1 Driving Situations

The driving task is often analysed in terms of three different performance levels provided by Rasmussen [8]: the knowledge-based, rule-based and skill-based levels. Differences between the levels relate to the involved mental effort. At the highest, i.e. knowledge-based, level, considerable attention and effort is required. At this level human behaviour is goal-controlled and represents a more advanced level of reasoning. Rule-based behaviour is characterised by the use of rules and procedures to select a course of actions. The rules can be acquired through experience or can be based upon prior instructions (training). When driving, rule-based behaviour involves interpreting everyday situations and applying rules and regulations that fit that situation. At the lowest, skill-based level, highly practiced tasks are carried out, requiring very little attention.

Rasmussen considers the amount of mental effort needed to execute a task and therewith addresses a dependency on individual differences in task performance. Michon [9], on the other hand, proposed that the driving task could be structured at three generic levels (independently from individual differences): the strategic, tactical and operational levels. At the strategic level drivers prepare their journey; this concerns general trip planning, choice of route, etc. At the tactical level drivers exercise manoeuvring control, allowing to negotiate the directly prevailing traffic circumstances, like crossing an intersection or avoiding obstacles. Here, drivers are mostly concerned with interacting with other traffic and the road system. The operational level involves the elementary tasks to manoeuvre the vehicle, mostly performed automatically and unconsciously (e.g. steering, using pedals or changing gears).

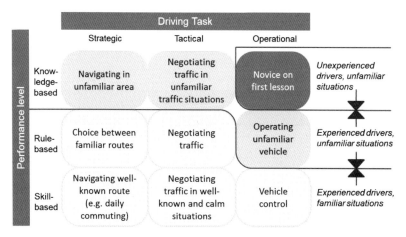

Fig. 2.   Traffic situations

Both models (the performance level taxonomy and driving task hierarchy) enable to classify driving tasks. Moreover, combining both models provides a good scheme to distinguish driving situations. The reason is that driving situations are characterized by environmental differences (e.g. road layout) in combination with individual differences of traffic participants (e.g. experience). This relation is very well recognizable in Fig. 2. An experienced driver would for example execute an operational task with his own vehicle at skill-based level, but might need knowledge-based performance for finding his way in an unfamiliar city.

3.2 Levels of Support for Partially Automated Driving

Before introducing an Assisted Driver Model, which has been composed to recommend driving support dependent on driving situations, we first need to explain which levels of intermediate automation should be distinguished. These levels have been derived from an existing taxonomy of automation-levels [7], called Levels of Automation (LOA). The reason why this taxonomy has been adopted is that LOA considers a scale of 10 intermediate support levels offered by partial automation of a task. These levels also cover the levels of automation theoretically possible for driving. Besides, LOA's aim is to facilitate appropriate system function allocations between human and computer controllers keeping both involved in the control loop –and this offers an important contribution to the avoidance of out-of-the-loop performance problems as indicated before. Levels of Automation (LOA) considers human and/or computer allocation to the following functions of the control loop: (a) Monitoring: Scanning displays or the system's environment to perceive information regarding system status and/or the ability to perform tasks, (b) Generating: Formulating options or strategies to achieve tasks, (c) Selecting: Deciding on a particular option or strategy, and (d) Implementing: Carrying out the chosen option. Based on LOA, we acknowledge 6 levels of support relevant for automated driving, which are indicated and explained in Table 1.

SUPPORT TYPES		FUNCTIONS				DESCRIPTION	EXAMPLES
		MON.	GEN.	SEL.	IMPL.		
1.	Augmenting	H/C	H	H	H	• Both human and machine monitor the present situation. The machine especially supports acquiring sensory information.	Night Vision
2.	Advising	H/C	H/C	H	H	• The machine supports by generating options, the human selects. The selected option might be another option than generated by the machine.	Attention Assist, Lane Change Assist
3.	Warning	H/C	C	C	H	• The machine temporarily generates *and* selects an option which, according to the machine, is mandatory to perform.	Lane Departure Warning, Frontal Collision Warning
4.	Intervention	H/C	C	C	C	• The machine temporarily generates, selects *and* executes an option which, according to the machine, is mandatory to perform.	
5.	Action Support	H	H	H	H/C	• The implementation part is being supported.	Powered Steering, Automated Gear Box
6.	Decision Support	H/C	H/C	H	H/C	• By combining Advising and Action Support, the human is being supported in terms of allowing full dedication to the selection-role.	
MON.= Monitoring task, GEN.= Generating options, SEL.= Selecting options, IMPL.= Implementation task							
H=Human task performance, C=Computer task performance, H/C= combined Human - Computer task performance							

Tab. 1. Indicating 6 levels of support relevant for semi-automated driving

3.3 Assisted Driver Model

The Assisted Driver Model [10] has been composed to recommend driving support dependent on driving situations. The model is shown in Figure 3. The previous section distinguished 6 intermediate levels of automation relevant for partially automated driving. To allocate these levels of automation to driving situations the Assisted Driver Model considered the prerequisites to provide good operation of the driving task. The considerations have been differentiated between the prerequisites for the involved performance level at one hand, and for the involved driving task type at the other hand. For the performance levels these prerequisites involve the avoidance of errors [8]. For the driving task types, the required level of perception and understanding (i.e. Situation Awareness) of the circumstances associated with the driving task, have been considered [3]. The selection of support types that fit both conditions resulted in the Assisted Driver Model.

3.4 Recommended Levels of Automation

The Assisted Driver Model shows for driving situations which are dominated by tactical and operational tasks executed at rule- or skill-based level, that automation is especially being recommended in terms of supporting the implementation task, i.e. Action Support. Within those conditions, Action Support enables the human to remain involved in task execution and preserves situation awareness, which allows better reaction times after failures and retrieval of control [11]. The model also advocates that driving situations which are dominated by option generating should not be supported in terms of joint human-machine task operation, i.e. Advising. Within those situations purely human generation of options performs far better than joint human-computer generation of options [7]. This superior human performance can be explained by distraction and doubts that humans encounter during joint human-computer selection of options.

Furthermore, the model shows that situations which require more intensive mental consideration (as is generally the case for strategic tasks) could be supported in terms of Advising. However, partially automation of decision making, like computer generation of options and human selecting, should be considered very carefully, for the same reason as mentioned before: Advising might cause worse performance due to doubts or confusion. However, because of the nature of these driving situations (i.e. strategic tasks which mainly involve way finding) alternatives are not available. With respect to performance after automation failure, tests show that recovery time is significantly lower with joint human-computer interaction during the implementation role, than with purely computer interaction [7]. This indicates that operator ability to recover from automation failures substantially improves with partially automation requiring some operator interaction in the implementation role.

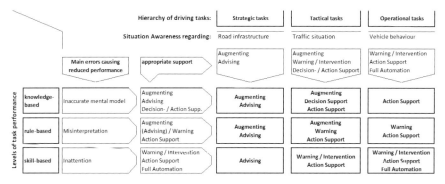

Fig. 3.   Assisted driver model showing recommended support types (i.e. levels of automation) dependent on driving situations

To summarize, the following levels of automation can be recommended in relation to different driving situations:

▶ Operational tasks benefit most from physical implementation assistance, requiring some human involvement. The human operator then remains involved in the control loop and this provides best recovery of control (after a transition from partially automation to full human control).

▶ Combinations of tactical and operational tasks performed at rule- or skill-based level benefit most from Action Support.

▶ Driving situations which are characterized by strategic tasks and/or dominated by option-generating are least appropriate for applying partial automation.

For some situations, it remains difficult to determine what level of automation is appropriate. We first notice a tactical task performed at knowledge-based level. This situation involves negotiating traffic in unfamiliar traffic situations and these circumstances typically involve decision-making, requiring considerable attention. Based on the model, either support in terms of Advising or in terms of Action Support would be recommended. Again, Advising could cause confusion. Action Support on the other hand could allow full dedication to the decision making part. Both types of support however, differentiate strongly upon the part within the control loop which is being supported. Therefore, further research is necessary to determine if and how partial automation would be beneficial for this situation. Also for an operational task performed at knowledge-based level it is difficult to determine what level of automation is beneficial. However, this situation involves novice drivers. Partially automation would therefore influence driving education and this is out of the scope for this research.

4   Final Comments

This research explained why large scale implementation of partially automated driving is more likely to become reality than completely automated driving. Based on human-centered considerations we identified two major challenges for the realization of partially automated driving: (1) Defining appropriate levels of automation, and; (2) Developing appropriate transitions between manually and automated driving. The Assisted Driver Model helped us with the first challenge, because the model recommends support types dependent on driving situations.

Next to when to provide automated driving, the question How to provide automated driving? is also important. The second challenge relates to this question. To develop appropriate transitions, a good starting point seems to review the possible levels of automation. As we have seen in chapter 3 especially support in terms of joint human-computer interaction during the implementation task, requiring some operator involvement, is recommendable. The reason is that with such support the human remains involved in the control-loop and therewith preserves awareness of the system status and surrounding traffic situation. An example is the implementation of pedals with force feedback. During automated cruising on a motorway (e.g. with Adaptive Cruise Control), the brake and acceleration pedals would continue to move or offer resistance to indicate the system's adaption in speed and distance in accordance with traffic situations. This would mean a more active involvement of the driver and allow better reactions when transitions to manual control are necessary.

Although support in terms of joint human-computer interaction during the implementation task allows better recovery, it will not necessarily make the driving task more comfortable. Examples from other area's (like aviation) often show that automation transforms human involvement from an operator-role to a supervision-role, without making the involved tasks easier, nor task performance safer. For the development of appropriate transitions in automation, it is therefore important to also acknowledge the relation with driver's acceptance. The fact that acceptance is more related to individual comfort, than advantages on a larger scale (like increasing traffic efficiency), leads us to a direction where we explicitly take performance of secondary tasks (e.g. listening to music or checking a dairy) into consideration. Interface solutions which combine performance levels for both the driving task and secondary tasks, could for example deliberately direct the driver's attention from a secondary task towards the driving task before automation terminates. However, future research, including experiments with simulated driving tasks, is required and foreseen to further develop appropriate interfaces for transitions between automation.

References

[1]   Van Arem, B., et. al., The impact of Cooperative Adaptive Cruise Control on traffic-flow characteristics, IEEE Transactions on Intelligent Transportation Systems, 7, no. 4, 2006.

[2]   European Commission, Raising awareness of ICT for smarter, safer and cleaner vehicles, Intelligent Car Initiative, Brussels, Belgium, 2006.

[3]   Martens, H. M., The failure to act upon critical information: where do things wrong? Doctoral Dissertation, Vrije Universiteit Amsterdam, 2007.

[4]   Hollnagel, E., A function-centered approach to joint driver-vehicle system design. Cognition, Technology & Work, 3, 169-173 (2006).

[5]   Martens, M.H. et. al. Human Factors' aspects in automated and semi-automated transport systems: State of the art. Deliverable no. 3.2.1. of CityMobil European project. European Commission, 2007.

[6]   Kaber, D. B. and Endsley, M. R.: Out-of-the-loop performance problems and the use of intermediate levels of automation for improved control system functioning and safety, Amer Inst Chemical Engineers, 1997.

[7]   Endsley, M. R. and Kaber, D. B., Level of automation effects on performance, situation awareness and workload in a dynamic control task, Ergonomics, 42 (3), 462-492 (1999).

[8]   Rasmussen, J., Skills, Rules and Knowledge: Signals, Signs and other Distinctions in Human Performance Models, IEEE Transactions on Systems, Man & Cybernatics, 13, 257-266 (1983).

[9]   Michon, J. A., A critical review of driver behavior models: what do we know, what should we do? In: Evans, L. and Schwing, R.C. (Ed.): Human behavior and traffic safety, pp. 485–520, 1985.

[10] Van den Beukel, A. P. and Van der Voort, M. C., An assisted driver model. Towards developing driver assistance systems by allocating support dependent on driving situations. Second European Conference on Human Centered Design for Intelligent Transport Systems, Berlin, 2010.

[11] Kaber, D. B. and Endsley, M. R.: Out-of-the-loop performance problems and the use of intermediate levels of automation for improved control system functioning and safety, Amer Inst Chemical Engineers, 1997.

Arie P. van den Beukel, Mascha C. van der Voort
University of Twente
Laboratory Design, Production and Management
Postbus 217
7500 AE Enschede
Netherlands
a.p.vandenbeukel@utwente.nl
m.c.vandervoort@utwente.nl

Keywords: driving support, driver assistance, ADAS, semi-automation, automated driving, levels of automation, transitions, driver interface, task allocation, Human Factors

Large-Scale Vehicle Routing Scenarios Based on Pollutant Emissions

D. Krajzewicz, P. Wagner, German Aerospace Center

Abstract

This paper describes simulation-based investigations of route choice based on pollutant emissions. A microscopic simulation enhanced by a pollutant emissions model was used to evaluate whether a vehicle's pollutant emissions can be used as an edge weight during route computation and which effects can be observed in such cases. For each of the pollutants CO, CO_2, NO_x, PM_x, and HC and for the fuel consumption, a dynamic user assignment has been performed. The investigations have been performed twice using two scenarios of different size. Large discrepancies for route computation using pollutants have been observed when comparing inner city and sub-urban traffic networks.

1 Introduction

The EC-funded project "iTETRIS" [1, 2] aimed at the development of an open source vehicle-to-vehicle/vehicle-to infrastructure (V2X) communication simulation environment. This environment was implemented by coupling two domain-specific simulators, ns3 [3] for communication simulation, and SUMO [4, 5] for traffic simulation, see also [6]. Aiming at being applied to current research topics, the work on accomplishing iTETRIS included the extension of both simulators. One of SUMO's extensions was the development and implementation of a model of vehicular pollutant emissions. The implemented model is based on HBEFA [7]. HBEFA was chosen from a set of 15 evaluated models due to having one of the largest coverage of vehicle population on the one hand, and allowing to model the pollutants' emission in dependence of the simulated vehicles speed and acceleration. The implemented pollutant emission model allows to compute the amount of emitted CO, CO_2, NO_x, PM_x, and HC and the fuel consumption for each vehicle in each simulation time step. For further processing, the computed emission values can be collected over the complete ride for each vehicle, or over variable time spans for each lane or road. A description of the emission model implementation can be found in [8].

iTETRIS took into concideration the ecological impacts of traffic by evaluating the influences of the developed V2X-applications on the amount of emitted pollutants, and by developing applications which use pollutant emission as an input. During the project execution, the interdependencies between pollutant emission and conventionally used values, such as travel time, delay time, or number of stops, were not clear. For this reason, it was decided to evaluate the characteristics of pollutant emission and the possibilities to use these measures for traffic management applications before trying to use them as an input variable for more complex V2X applications. Traffic signal control and the route choice have been chosen for evaluation, since they are often considered and adjusted in traffic management methods. The initial results of both evaluations are given in [9] and show a strong relationship between conventionally used values, such as the waiting time or the delay time, and pollutant emission.

In the following, further insights into the routing of vehicles using pollutant emission are presented. First the dynamic user assignment (DUA) paradigm and the method used within the described investigations for achieving a DUA are described. The scenarios and the results of the simulation of route choice based on pollutant emissions are then given. The document ends with a discussion.

2   Route Choice Paradigm and Algorithm

It is known that a street's or road's throughput or traffic flow highly depends on the number of vehicles on that road. For lower densities (k), traffic flow (q) is increasing with an increasing number of vehicles that use this street. At a saturation point however, the traffic flow breaks and decreases with an increasing number of vehicles. Being one of the main traffic characteristics, its visual representation is known as "the fundamental diagram" of traffic. One example using data collected from the A9 highway in Germany is given in figure 1. This traffic behaviour is explained by an increasing interaction between vehicles, and the disability of drivers to take constant headways, see [10] and [11]. The major consequence arising from this characteristic is that the travel time on a road changes with the number of vehicles on it.

Road travel times depend not only on the number of vehicles running within a network but also on the capacities at intersections. As soon as the number of vehicles approaching an intersection is getting greater than the intersection's capacity, growing queues will arise in front of the intersection. Its respective travel time will then increase. At signal-controlled intersections, new arriving vehicles may then have to wait for more than one cycle.

As a conclusion, when computing a fastest way through a road graph, one has take into account the travel times under a changing load of the graph's edges. Dynamic user assignment denotes a process of computing routes for a given traffic demand and take into account the change of road weights over time. Such assignment is, besides preparing the network and the demand data, one of the major steps for setting up the traffic simulation for a certain area.

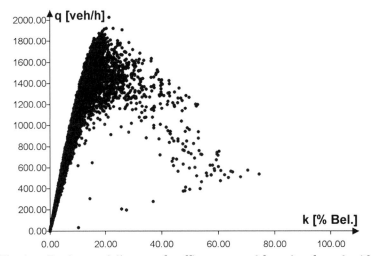

Fig. 1.   Fundamental diagram of traffic - generated from data from the A9 highway in Germany

The algorithm of the adopted traffic assignment model was developed by Christian Gawron [12] and allows to compute a dynamic user assignment with a given road network and a given demand where the latter is given as a set of vehicular routes. Each route is defined at least with a vehicle's departure time and a vehicle's start and end road. It is a microscopic algorithm, computing the route to use for each of the modelled participants individually. The algorithm was shown to achieve a dynamic user equilibrium (DUE) where each participant has the fastest possible route given the fact that he is not the only road user. It should be emphasized that the resultant distribution of routes over the network does not constitute a system optimum. Other methods for assignment could yield in an overall - system-wide - reduction of route durations' sum, see [13], but this normally is accompanied by longer routes for a portion of the participants.

The Gawron algorithm used here is an iterative algorithm, where route choice and simulation are performed subsequently until a desired route distribution state is reached. Within the first step, all vehicles are assigned to the fastest routes through the empty network, as no other information is available at this time. Then, the simulation is executed using these routes. The simulation outputs average travel times for each edge of the simulated road network graph. The travel times are normally saved as a time line of values for each edge, as they change over time. Often, a time aggregation of 15 minutes is used. In subsequent steps, the travel time information obtained from the last simulation run is read by the routing module. The computation of best - normally fastest - routes may now yield in a new route, as the edges' travel times have changed due to using the weights from the last simulation run. In order to avoid oscillations, Gawron's algorithm chooses one of the so obtained route variants randomly, weighted by the routes' benefit (travel time) in comparison to the other variants. The algorithm by Gawron has proved to be reliable during our past work on traffic management. Descriptions of the work on improving the algorithm's speed, and on comparing it against other, better known algorithms, can be found in [14].

For the evaluations reported here, the routing application was extended by the possibility to use two time lines of values. The first time line contains the travel times for each edge as described above. The second time line contains emission values for each edge and can be used during the routing step just like the travel times.

3   Simulation and Results

Two scenarios were used for the investigations. Both were generated within the iTETRIS project and model the roads within the city of Bologna and its surroundings. The first scenario includes the area around the middle-age inner city ring. The second one is based on the first scenario and additionally includes the highways and rural roads around the inner-city centre. Traffic light programs from the real world were embedded in the scenarios and both scenarios include the traffic demand for the morning peak hour between 8:00 and 9:00. Figure 2 shows the road networks of both scenarios. A more detailed description of the scenarios can be found in [10].

The investigations were done by computing a user assignment with use of the amount of fuel consumption and the emissions computable by the simulation – CO, CO_2, NO_x, PM_x, and HC – as edge weights instead of the normally used travel time. This was done separately for both presented scenarios. In both cases, a conventional assignment using travel times was computed as a baseline for comparing the performance of emission/consumption-based assignments. For each pollutant and the travel time, the assignment loop was repeated 50 times and 30 times for the small and the large scenario respectively. After each assignment, the simulation was re-run with use of the routes obtained from the last iteration run. During this run, the simulation was set-up to write measures summed up over roads, and over vehicles, separately. In the following, the per vehicle values which include both conventional measures, such as the vehicle's route length, travel time, total waiting time, and ecological measures, consisting of the amount of emitted pollutants and fuel consumption during the vehicle's complete ride are used.

Fig. 2.   Used scenarios; left: the smaller "ringway" scenario covering the inner-city centre, right: the larger "complete" scenario

Figure 3 shows the deviations of the measures collected from pollutant emission / fuel consumption-based assignments from those obtained by using the routes computed by a travel time-based assignment. It should be noted, first, that using pollutant emission / fuel consumption for computing routes that minimize the respective measures does work. In most cases, the optimized measure's value is lower than in runs which optimized a different one. Within the smaller "ringway" scenario, optimizing routes towards a pollutant's emission reduction also yields in a route set which generates less of the other pollutants. This is no longer the case for the larger "complete" scenario.

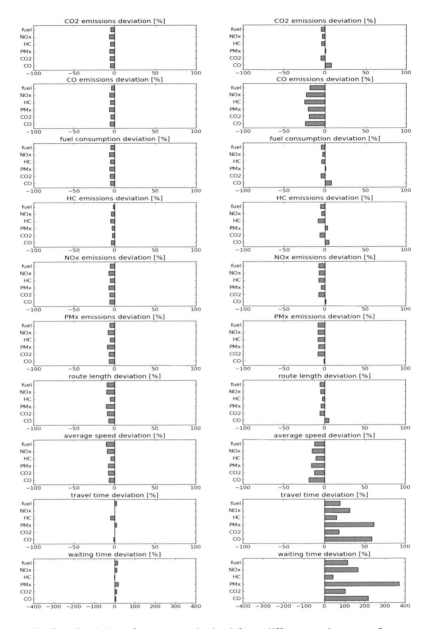

Fig. 3. Deviation of measures obtained from different assignments from the assignment based on travel times. Left column: "ringway" scenario, right column: "complete" scenario. The labels on the y-axis denote the measure against which routes were optimized. Please note the scale change in the last row.

For example, routes optimized against CO or PM$_x$-emission result in a higher average fuel consumption and have a higher HC emission value than those obtained from travel time-based assignment. Emission-optimized routes are in average shorter than those obtained from the travel time-based assignment. Being passed at a lower speed, they still do not have a lower travel time. Within the small scenario, the average travel time stays similar along all assignments. However, within the large scenario, the emission/consumption-optimising routes have a much higher travel time than routes computed using the travel time as weight. The reason for this is a large - up to almost 400% - increase in waiting times compared to travel-time based assignment. This appears plausible as a standing vehicle emits few pollutants but incurs a direct increase in travel time. The outputs generated by the simulation allow to compare the number of vehicles that pass the simulated network's roads. For the smaller "ringway" scenario, only few distinctions were found - the number of vehicles per edge is almost the same for all performed assignments. However for the large scenario clear shifts in road usage can be observed. This is especially the case when optimizing against CO emission. Figure 4 shows the shift of routes from rural roads to highways by showing the difference in road usage when optimizing CO emissions instead of travel time.

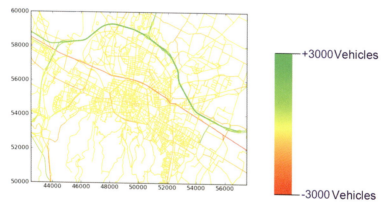

Fig. 4.   Difference between road usage

In order to determine how conventional and ecological measures correlate, we computed the Pearson-correlation between the collected values on a per-vehicle base. Here, only the values from the iterations which used travel time as edge weights are used. The results are given as figure 5. They show a strong correlation between all pollutants' emission, and a slightly lower, but still high correlation between the routes' durations and the amounts of emitted pollutants. A topic for further investigation is the change of the measures' "waiting time" and "average speed" correlation with emitted pollutants.

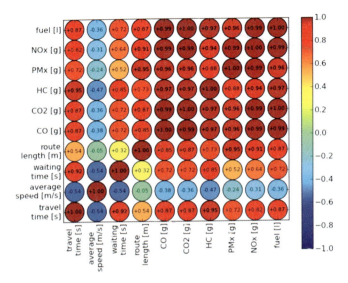

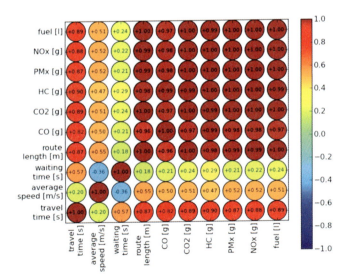

Fig. 5. Pearson-correlation between collected values; top: "ringway", below: "complete"

4   Conclusion

Investigations on computing routes for vehicles using other measures than the usually used travel time have been presented. It was shown, that optimizing routes towards minimization of a certain pollutant or fuel consumption is possible. The observed optimization seem to depend on the type of roads available in the investigated area.

The results can be applied to develop methods for reducing the network-wide emission of a certain pollutant by assigning vehicles new routes. Before being applicable, further investigations on the inter-dependencies between pollutant emission and road network types must be performed.

References

[1]   iTETRIS web-site, http://www.ict-itetris.eu/, 2011.

[2]   Lázaro, O., Roads to the Future, ITS Traffic Technology International Magazine, October/November 2010.

[3]   ns3 web-site, http://www.nsnam.org/, last visited on 13.03.2011.

[4]   SUMO web-site, http://sumo.sourceforge.net/, last visited on 13.03.2011.

[5]   Behrisch, M., Erdmann, J., Krajzewicz, D., Adding intermodality to the microscopic simulation package SUMO, 2010, Alexandria, Egypt.

[6]   Maneros, J., Rondinone, M., Gonzalez, A., Bauza, R., Krajzewicz, D., iTETRIS Platform Architecture for the Integration of Cooperative Traffic and Wireless Simulations, ITS-T 2009, Lille, 2009.

[7]   INFRAS, Handbuch der Emissionsfaktoren, http://www.hbefa.net/, 2011.

[8]   iTETRIS Consortium, Deliverable D3.1: Traffic Modelling: Environmental Factors, February 2009.

[9]   iTETRIS Consortium, Deliverable D3.2: Traffic Modelling: ITS Algorithms, April 2010.

[10]   Krauß, S., Microscopic Modelling of Traffic Flow: Investigation of Collision Free Vehicle Dynamics, Hauptabteilung Mobilität und Systemtechnik des DLR Köln, 1998.

[11]   Brilon, W, et al, Fortentwicklung und Bereitstellung eines bundeseinheitlichen Simulationsmodells für Bundesautobahnen, Forschung Straßenbau und Straßenverkehrstechnik, Heft 918, Herausgegeben vom Bundesministerium für Verkehr, Bau- und Wohnungswesen, 2005.

[12]   Gawron, C., Simulation-based traffic assignment – computing user equilibria in large street networks, PhD thesis, Universität zu Köln, 1998.

[13] Wang, Y.-P., Wagner, P., Behrisch, M., Towards a dynamic system optimum based on the simulated traffic data in the microscopic traffic simulation, 3rd NEARCTIS workshop, Schweiz, 2010.

[14] Behrisch, M., Krajzewicz, D., Wagner, P., Wang, Y.-P., Comparison of Methods for Increasing the Performance of a DUA Computation,. Proceedings of DTA2008 International Symposium on Dynamic Traffic Assignment, Leuven, 2008.

Daniel Krajzewicz, Peter Wagner

German Aerospace Center
Institute of Transportation Systems
Rutherfordstraße 2
12489 Berlin
Germany
daniel.krajzewicz@dlr.de
peter.wagner@dlr.de

Keywords: pollutant emission, traffic management, route choice, green mobility

EV-Cockpit – Mobile Personal Travel Assistance for Electric Vehicles

J. C. Ferreira, Institute of Engineering of Lisbon
A. R. da Silva, IST/INESC-ID
J. L. Afonso, University of Minho

Abstract

This paper describes the concept and a prototype proposal for a mobile information system to support the mobility process in cities, such as it gives recommendations to drivers about public transportation, car or bike sharing systems. Electric vehicles (EVs) integration will contribute to the reduction of usage of private cars in the cities and also to decrease the CO_2 emissions. EVs will play an important role in the integration of renewable energy sources (intermittent sources) in the open electrical market, and the complexity of the charging process will be a great opportunity for the development of systems oriented to mobile devices and to social networks. The main objective of this proposal is the creation of a platform based on successful approaches developed in the computer science area, like recommender systems, cooperative systems and social networks, to help the creation and establishment of smart cities.

1   Introduction

The automobile is playing an important role in transportation and new challenges appears with the introduction of the electric vehicle (EV): The so-called city electric vehicles that are EVs with an autonomy of less than 160 km. Drivers of such EVs will have to cope with a new problem: EV charging takes time and they need to deal with limited electric range - a problem referred to as "range anxiety" [1]. The success of EV in part is going to depend on how comfortable people feel with getting where they want to go, without running out of charge, and without having to go through some process that will take them a long time and impact their ability to use the vehicle. So, taking out in real time EV information, such as SOC (State of Charge), energy transactions and others vehicles events plays an important role. This information, taking into account recent progress in mobile devices, geographic information system and communication processes, can bring added value to drivers.

EV-Cockpit is a mobile system for EVs drivers that bring the 'right' information to them. With regard to prospective developments in smart grids, open energy market, and smart cities with increasing mobility sustainability, the proposed system integrates a diversity of functionalities, as illustrated in Figure 1.:

▶ ICT Supported Functionalities: The EV-Cockpit System receives the geographical information of the EV current position and the features that enable the calculation of distances between two points. Main ICT (Information and Communication Technologies) contributions are the GPS devices, mobile devices and wireless communication for user's information access from anywhere, and also the corresponding geographic information system.

▶ Mobility Sustainability Functions that involve the following: (1) Car sharing and bike sharing, systems to support mobility functions in 'smart' cities [2, 3]. (2) Integration of Public transportation information to create a route planner based on public transportation. (3) Information on Points of Interest: Information about points of interest is preloaded on the system and is used for direct consultation by the driver, who can perform a quick search for points of interest near the present location. The information is also used in the recommendation of charging points, for all that remain at a distance, for example less than 5 km, will be marked as being "near the point of interest" [5]. (4) Parking spots available and remote reservation. (5) Route planer, based on the integration from different data sources, such as multi-modal public transportation systems, car and bike sharing, or car pooling [3].

▶ Energy Market Functions: (1) Aggregator for energy market participation. (2) Collaborative broker for Distributed Energy Resources (DER) [6]. (3) Account system for electricity transitions with price control. The EV-Cockpit system receives the discounted value of the price of energy from the energy market, consulting it regularly. The only information received is the fare in the current format €/kWh. The energy market information is used to control the battery system during periods of loading. If the price of energy rises above a configurable threshold, the EV-Cockpit system sends a command to the battery system to stop charging. If the price falls back to an acceptable level, an alert is sent to start charging again.

▶ Charging Functions that involves the following: (1) Range anxiety utility functions, such as: display of SOC, remaining kms for next charging, driver profile, looking for nearest charging station (with guidance and charging spot reservation). (2) Remote charging control interactions, by the possibility of remote commands. (3) Tracking system. (4) Consumption simulation for smart charging strategies taking into account distribution limitations and establishing of a smart charging strategy. (5) Micro grid integration.

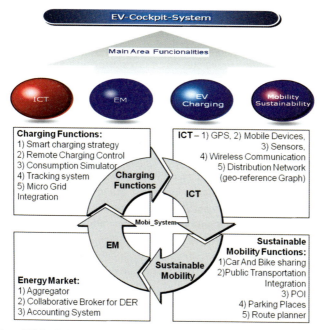

Fig. 1.   EV-Cockpit main functionalities and purposes

2   Mobility Sustainability and Energy Market Functions

Our approach considers a central system with the aim of creating conditions and incentives for drivers to use less their own car, by giving guidance and suggestions for others transportation systems, like public transportation, bike sharing, car sharing or even car pooling [2]. EV-Cockpit system works also based on real traffic information, supporting decision for the best options taking into account pre-defined criteria (e.g., fastest option, cheapest option, option with less CO_2 emissions), as Figure 2 suggests.

EV-Cockpit is a system [3] integrated in the START European project [4], that allows citizens to obtain information (e.g., timetables, routes and prices) on the various modes of transport (e.g. bus, tram, metro, train, ferry) available in a particular region or city, focusing on the integrated use of soft transport (e.g. electric vehicle, bicycle) and occupation of waiting time (eg, visiting points of interest), and is based on a local domain ontology for public transportation data integration and others systems, like bike and car sharing, or even car pooling. The idea is to pass public transportation options to a graph, where the arch length is defined by the time that takes to go from one node arch to the other.

The same procedure is applied for car sharing, car pooling and bike sharing systems. If we have all this information in a graph, we can use the Best Path algorithm (a Dijkstra algorithm implementation) [3], where the arch weight that connects to adjacent nodes can be constructed from a diversity of options, like related time, price and CO_2 emissions.

The main contributions of this work are the following: (1) a domain ontology definition for Public Transportation; (2) a data integration of Public Transportation in European environment; (3) a functional prototype to query and give advice regarding the best path from a point A to point B using Public Transportation facilities; (4) an implementation in RDF/RDFS (Resource Description Framework Schema) that allows the use of the emergent query language SPARQL; and (5) an implementation of the best path's Dijkstra algorithm and a diversity of public transportation information sources. Details for this can be found in [3, 7]).

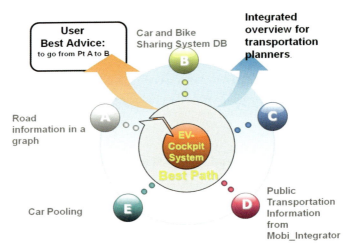

Fig. 2.   EV-Cockpit System transportation related system

Regarding energy market functions, because of their batteries, EVs present an interesting potential as a storage facility. However, the storage capability of the EV batteries is small on the grid scale, and consequently their individual power output cannot have any impact on the power system. For the EVs to be able to play a role when interconnected to the grid, they need to be grouped into communities. Once aggregated, they are able to provide different kinds of services, either as a controllable load or as a generation/storage device. However, the EVs may not be always plugged into the grid and their schedules are very uncertain.

We have developed a conceptual V2G (Vehicle to Grid) aggregation platform to solve this aggregation problem, based on a collaborative approach with a credit mechanism to measure user participation and to divide energy market participation revenue [8]. Also, other aspect of this problem is the integration with micro-generation and the problem of handling the distribution energy sources [7]. In this process is important to determine the EV driver behavior through user profile with information like time of trip, distance, daily hours connected to the grid, or minimum energy stored. For that we have developed a tracking system to work offline in a mobile device with GPS [9, 10].

3   Charging Functions and Users Applications

Main charging functionalities related to the charging process and to dealing with the range anxiety problem, which are illustrated in Figure 3. One major issue in this problem is the vehicle external information access and sharing among the different stakeholders. Vehicle manufactures traditionally block most vehicle external information access, but community will benefit from a collaborative information sharing. There is a progressive tendency for the creation of laws to oblige OEM (Original Equipment Manufacturers) to share vehicle information, but we will not discuss this here. But once this issue is solved (that information should be shared), there are two approaches: (1) OEM starts this business and transmits related information to nearby mobile devices through Bluetooth; or (2) through the usage of a standard communication interface (e.g., CAN-Bus interface) the relevant information is extracted in real time. We have explored this second approach to develop an OBU (On Board Unit), and have performed some work on it, based on a microcontroller that integrates CAN, Bluetooth, GSM/GPRS (global system for mobile communications / general packet radio service) and GPS (global positioning system). The implementation of CAN protocol allows to receive real time data from EV. With the available OBU wireless communications interfaces, it will be possible to report both locally and remotely the data being received from the EV through Bluetooth and or GSM/GPRS technologies, respectively. Moreover, Bluetooth allows the OBU integration with mobile equipment, such as a mobile/smart phone. Additionally, and by having knowledge of the EV current coordinates (GPS receiver), the OBU will be able to make the best decisions through the platform. GPRS will allow the development and implementation of the OBU update over the air, increasing the easiness with which software updates are made.

CAN is a vehicle bus standard designed to allow microcontrollers and devices to communicate with each other within a vehicle without a host computer. Thus, the idea is to develop ways of taking vehicle relevant information to computer systems (mainly mobile devices). This allows feeding our developed application with real data from EV. From this case study, since there are standards, the idea is to create an open tool that can work in all vehicles with CAN-bus. User information access is mainly performed by a desktop application (V2G Smart System) or by a Mobile Application.

The main functionalities of the V2G Smart System web application are: (1) Registration: registration page for new users; (2) Password Recover: form for password recovery; (3) Login: home page of the application - the user is redirected to this page after login; (4) Profile Creation: page created for user profile by entering information on the EV; (5) Personalized Charge Profile: page load profiling, through the introduction of information regarding the date / time of travel, number of km the driver intends to accomplish, and minimum SOC allowed for the EV batteries; (6) Statistics: home energy consumptions, weekly, monthly and annual energy expenses, price variation of electricity, charging periods, etc. The present application on the server is subdivided into five main modules: (1) Interpreter of Downloaded Files - this module will be responsible for reading and interpreting the files loading, giving the system a layer of abstraction over the file format of text issued by the loading system; (2) Smart Grid Interface - this module will be responsible for the interaction with the electrical network, i.e., it controls the flow of energy from or to the electrical network, with the objectives of helping network stability, and also, managing information on the variation of electricity prices, to optimize the profits obtained with the selling of energy to the electrical network; (3) User Manager - module responsible for registering the users and their EVs, allowing the recording and editing of users data, as well as the removal of users (if defined rules are not accomplished by specific users) - this module is also responsible for verification of user identity and ownership of registered vehicles (through the transmission of data received from the user to the authorities), and for performing regular cleaning from the database of users categorized as "spam"; (4) Manager Profiles - a user can set one or more load profiles for each of the vehicles registered by him. A common practice is, for example, the definition of profiles and needs of different charging to be carried out during the week (weekdays) over the weekend; and (5) Manager Central - interacting with various modules mentioned above, and managing the distribution of system information (from other modules and database). A Central Information Repository will store EV related information, namely: EV drivers profile, electricity transactions of EV, electricity prices, and other EV related information. Details of this system can be found at [8, 9].

The Mobile Application, called EV-Cockpit-Mobile application is used by vehicle drivers to interact and take useful information from the EV itself, but also from charging devices, the Energy Market and the central information server. Due to the diversity of mobile operator systems, this application is developed on top of Android and Windows Mobile platforms.

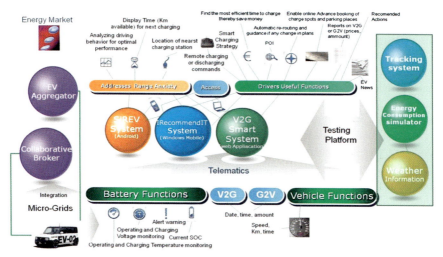

Fig. 3.   EV-Cockpit functionalities

The driver of the EV can use the EV-Cockpit-Mobile application to: (1) get directions; (2) locate and load charge spots data, reserve slots; (3) get recommendations on his journey with great information needs, as local battery charging, whose constraints are known involving the loading time and their autonomy, local information satisfaction, public transport information in case of failure, etc; (4) get points of interest; (5) define smart charging strategy; and (6) define parameters related to electric market participation (e.g., selling, buying and profit maximization).

The key functional requisites for the EV-Cockpit-Mobile application, illustrated in Figure 3, are the following: Firstly, a process to manage charge/discharge, where the charger or discharger device should receive relatively simple commands such as: charge the battery, wait for further instructions, return energy to the electrical grid, and archive transactions for further analyses. Secondly, the driver mobile device with GPS should track the EV movements in an offline mode (avoiding user charges of GPRS connection). Thirdly, guidance for charging stations, their location, and the reservation of charging spots. Fourthly, the user should establish a profile (stored on user's PC) where he defines his habits (e.g., number and time of travels per week, travel time and distance, weekend habits), and also the minimum SOC level (that allows to drive for the minimum distance). Fifthly, energy market functions, such as to obtain the price information to sell energy. Sixthly, a tracking application should be installed on the mobile device and configured to work in offline mode (to avoid charging expenses). A prototype for Android [5] and Windows Mobile [10] operating systems have been developed.

3.1 Charging Platform – IV2G System

A charging device was developed based on a mobile device, in order to perform remote commands and to establish a smart charging strategy, taking into account energy price and power limitation. In typical electric vehicles, when it is necessary to charge the batteries, the energy comes from the electrical grid to the batteries in unidirectional mode, without any control protocol given by the electrical grid. For more details, please see [10].

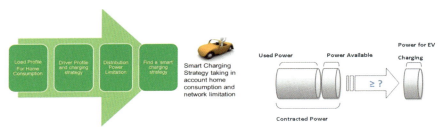

Fig. 4.   Smart Charging approach and goals

3.2 Simulation of Home Consumption and Finding of Smart Charging Strategy

An agent based electricity consumption simulator was developed [9, 11] that allows determining the best EV charging process taking into account home and distribution power limitation. Furthermore, a study was made for different types of residential consumptions with the goal to analyze the introduction of the electric vehicle. It took into account the different profiles of families (with different power consumptions and traveled distances), assuming one vehicle per family, and determining the most appropriate forms to charging the EV. Then, it was determined the time of day when there is a larger amount of energy to be used, and consequently, compiling these values of consumption per hour it was found the ideal intervals along the days for EV charging. Also, this platform can be used for the simulation of real testing environments of EV charging, adapted to different countries specificities. Also electric distribution companies can use this tool for future planning, simulation and decision support. Consequently, this information can be used to determine the capability of the actual electric distribution network for supply energy to the final consumers, and also for charging the EVs banks of batteries, which can occur simultaneously. With this tool we are able to determine a Smart Charging System in order to achieve the goals as identified in Figure 4, i.e., taking into account home consumption and electrical network power distribution limitations, the proposed system identifies a smart charging strategy.

4   Conclusion

The main goal of this work is to bring ICT and Information System approaches to this upcoming growing area of sustainable mobility process in smart cities, with the introduction of EV and Energy Market participation. All proposed system modules are oriented in a way that mobile devices should be part of drivers' mobility process.

In EV charging and discharging processes, a management system is created with a smart charging strategy taking into account distribution and consumer power constraints. Since there is no real environment for testing purposes, a simulation tool based on agents was developed. Based on this tool and taking in consideration real information (user's surveys and information taken from tracking system) a home consumption analysis was performed based on an agent using a stochastic process. Taking into account the limitations of the distribution network and the user's power contracts, a smart charging strategy was identified (basically this is an indication of maximum power through the time that the charging process can use). In our opinion more complexity can be introduced on this tool, such as microgeneration and discharging process. The geo-reference in a graph, the electrical distribution network and a visualization tool are also proposals, which could help the planning of new distribution infrastructures, and the identification of regions of power constraints.

Considering the Energy Market participation, the main proposal is a conceptual system to create and manage the EV community, with a credit-based approach which is an innovative proposal of this work, together with the collaborative broker for Distributed Energy Resources (DER). In our view, using this credit-base system, together with rankings, the users would profit from an open and healthy competitive environment. Also, in the future, to increase the market share of EVs, there will be a need for these types of systems in order to explore the potential of the energy market for these kinds of vehicles. Also, renewable energy sources integration and microgeneration can benefit from a community coordination action, where users capture renewable energy produced in excess at lower prices at local generation, and also avoiding transportation loses.

In transportation, a diversity of systems to reduce usage of own vehicles is created, reusing pieces of software components, and with approaches created for charging process and Energy Market participation (credit mechanisms, charging spots reservation). The main contribution is the public transportation data integration and the diversity of systems (car sharing, bike sharing, car pooling) available to the users through a single interface: the EV-Cockpit system. This holistic view is important for EV industry in general, but also for end-users, for governments and for public operators, because it gives an integrated overview, where it is easy to identify the challenges and the opportunities of this area.

References

[1] http://www.thebigmoney.com/blogs/shifting-gears/2010/05/06/survey-says-electric-car-range-anxiety-real

[2] Ferreira, J. C., et. al., Collaborative Car Pooling System Proceedings of Int. Conf. on Sustainable Urban Transport and Environment, Paris, 24-26 June 2009.

[3] Ferreira, J. C., Afonso J. L., Mobi_System: A Personal Travel Assistance for Electrical Vehicles in Smart Cities. 20th IEEE International Symposium on Industrial Electronics (ISIES 2011), Gdansk University of Technology, Poland, 2011.

[4] START Project: http://www.start-project.eu/en/Objectives.aspx

[5] Ferreira, J. C., et. al., "Recommender System for Drivers of Electric Vehicles". International Conference on Network and Computer Science (ICNCS 2011), Kanyakumari, India, 2011.

[6] Ferreira, J. C., et. al., Collaborative Broker for Distribuited Information Sources. 9th IEEE International Conference on Industrial Informatics (INDIN2011), Lisbon, Portugal, 2011.

[7] Fernandes, T., Fontes, M., Determinação do melhor caminho em sistemas de transporte terrestre (BPath – Best Path). ADEETC – Final year project (ISEL – 2009).

[8] Ferreira, J. C., Afonso, J. L., "A Conceptual V2G Aggregation Platform", EVS-25, Shenzhen, China, 2010.

[9] Ferreira, J. C., et. al., Simulation Platform for Electric Vehicle charging Process. 1st International Electric Vehicle Tecnology Conference (EVTEC 11), Yokohama-Japan, 2011.

[10] Ferreira, J. C., Afonso, J. L., "Towards a Collective Knowledge for a Smart Electric Vehicle Charging Strategy". International Conference on Industrial and Intelligent Information (ICIII 2011), Bali, 2011.

[11] Monteiro, V., et. al., "iV2G Charging Platform", IEEE-ITSC, 13th International IEEE Conference on Intelligent Transportation Systems, Madeira, Portugal, 2010.

Joao Carlos Ferreira
Institute of Engineering of Lisbon
Rua Conselheiro Emídio Navarro 1
1900-049 Lisboa
Portugal
jferreira@deetc.isel.ipl.pt

Alberto Rodrigues da Silva
IST/INESC-ID
Rua Alves Redol 9
1000-029 Lisboa
Portugal
alberto.silva@acm.org

João Luiz Afonso
DEI – University of Minho
Campus de Azurém
4800-058 Guimarães
Portugal
e-mail: jla@dei.uminho.pt

Keywords: electric vehicle, charging system, energy market, transportation data integration, community, best path, mobile device, distributed energy resources

DTF - A Simulation Environment for Communication Network Architecture Design of the Next Generation of Electric Cars

A. Hanzlik, E. Kristen, AIT Austrian Institute of Technology GmbH

Abstract

The Data Time Flow Simulator DTF is a discrete-event simulation environment that is currently under development at the AIT Austrian Institute of Technology GmbH. The DTF is developed in the context of the POLLUX project that is related to the control electronics architecture design of the next generation of electric cars.

1   Introduction

Like the conventional automotive industry, the market for electric cars can be expected to become a market of mass production. On one hand, due to the fact that the same platform may be used for different generations of electric cars, the lifecycle of a communication network architecture (physical network and communication schedule) is expected to be 20 years or longer. On the other hand, future extensions to the car functionality come along with bandwidth requirements. The technical challenge is the design of a communication architecture that fulfils existing requirements and that is open for new functionalities without the necessity to re-design the whole communication network. Evidence from the automotive industry shows that the design of such a network architecture is an arduous task and may be an issue of development and test for several months. Once defined, the communication architecture shall be valid for the lifecycle of the product. Additionally, there are safety requirements to fulfil, that have to be reflected in the design of the communication architecture. Overall, a design tool for development and validation of communication network architecture seems to be a desirable and valuable support during the development process of an electric car.

This challenge is addressed by the DTF Data Time flow simulator, a discrete-event simulation environment that shall help to identify and validate optimal car communication network architectures with regard to technical requirements, safety requirements and industrial norms. The idea is to derive the communication architecture in a very early design phase, e.g. from the requirements specification, to build a simulation model of the architecture and to test the communication network in the simulation environment. If the tests reveal requirements violations (e.g. deadlines of safety-relevant signals are missed), the communication network architecture is re-designed in the DTF simulation environment or the communication schedule is modified. This cycle is repeated until all requirements are met.

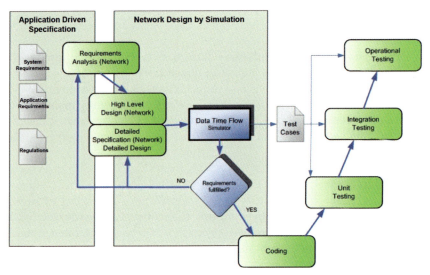

Fig. 1.   DTF Development Cycle

2   The DTF Simulation Environment

Figure 2 shows the structure of the DTF simulation environment. Starting with a graphical description of the system, the converter generates a textual bus model description that in turn is the input for the parser. The parser checks the bus model description for correctness and consistence and passes the model description to the DTF simulator engine. The DTF simulator contains no error-checking mechanisms with regard to the model description; this is the task of the parser. The DTF simulator can therefore rely on a correct model description. The simulator engine creates the necessary elements according to the model description. Then, the event list is generated according to the drive cycle description. An example for a drive cycle description is a sequence of events that defines different accelerator pedal positions for different points in time. Last, the event list processing is started with the first event. In the course of the simulation process, an event received at an element may lead to the creation of another event that is added to the event list. Simulation is complete when the event list is empty or the simulation end time is reached. All events are logged into a simulation results file when simulation is complete. This file can be processed offline to gain data necessary for visualizations and further analysis.

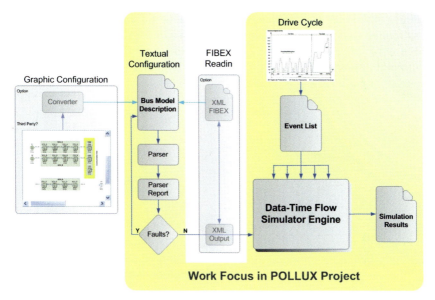

Fig. 2.   DTF simulation environment

2.1 Principle of Operation

The DTF principle of operation is to build up complex communication systems from primitive building blocks (e.g. sensors, processors, actuators), the elements. Each element has an input buffer, a delay and an action routine. Upon reception of an event, the associated element reads its input buffer, generates an output value according to its action routine, and transmits the output value with a defined delay to the next attached element in the action chain. Elements are activated by event triggers or time triggers. An event trigger is set by the predecessor element and a time trigger is set periodically by the element itself. Each element has one or more inputs and one output.

2.2 Simulation Procedure

Signals are issued to the sensor elements and signal propagation is observed, both in the domains of value and time, from the sensors over the communication network to the actuators. From the signal propagation time distribution over the communication network important information can be gained with respect to the dynamics and responsiveness of the system, especially for safety-relevant signals like the accelerator pedal position that usually are subject to real-time constraints. Additionally, the communication network can be exposed to stress via so-called disturbance nodes to assess the impact on the timeliness of safety-relevant signals. The amount of traffic generated by the disturbance nodes, until a deadline of a safety-relevant signal is missed, delivers expressive information about the stability reserve, robustness and extensibility of the communication network architecture.

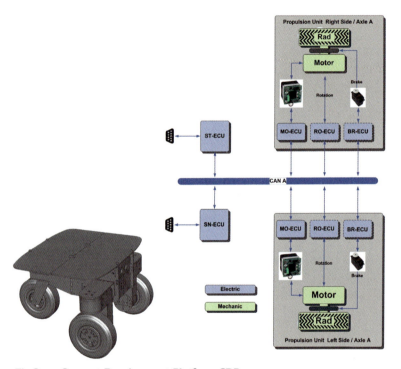

Fig.3.　　Concept Development Platform CDP

3　Validation

For validation of the DTF implementation a Concept Development Platform (CDP) is used to compare simulated results and bus measurements from the CDP to optimize the DTF models. The CDP supports the architecture of a vehicle network system in a versatile and easy to handle format. The platform provides the mechanical system to develop automatic control software for the powertrain as well as the necessary structures for network design.

3.1 CDP Structure

Figure 3 shows the CDP. It is a single-axle driving platform where each wheel is driven by one electromotor, stabilized by a free-running support wheel at the back. If both wheels run with the same speed, the CDP drives straight ahead. To go left, the right wheel has to be accelerated (or the left wheel has to be decelerated or both). In reverse, the same applies for going right. Larger speed differences between the two wheels result in closer curve radii driven by the CDP.

The CDP is directed via a remote control that sends driving commands to a Command Electronic Control Unit. The driving commands correspond to the desired nominal rotation speed for each wheel. The Command ECU has two nominal speed command sensors (inputs), one for each wheel. Each sensor receives the nominal speed value from the remote control. With a period of 10ms, the two values are packed into a CAN frame that is sent over the CAN A network to the Motor ECU's where the CAN frame is stored. Every 20ms, each Motor ECU extracts the nominal speed value for its electromotor and forwards this value to a control algorithm that calculates a control variable for the motor power driver. Each motor output is attached to an actual speed sensor (input) of a Rotation ECU that receives the actual speed value of the rotational speed of the motor. The actual speed value is packed into a CAN frame and is sent over the CAN network to the corresponding Motor ECU. In general this forms a closed-loop control system within the distributed control architecture. From now on, the control algorithm updates the motor control variable according to the nominal value and the actual value with a sampling rate of 20ms.

3.2 Simulation Model

For the validation experiments, we use a DTF simulation model of the CDP architecture shown in Figure 3. Unsurprisingly, the structure of the simulation model closely reflects the CDP structure outlined in the last section.

The principal super-elements in Figure 5 are the CAN Bus (in the middle), the CMD_ECU (left-hand side, top) and the Motor ECU's MO1_ECU/1 and MO2_ECU/1 (right-hand side, respectively. The two electro motors can be found at the bottom of Figure 5. The CAN Bus is responsible for communication between the different super-elements. The CMD_ECU receives driving commands from a remote control and transmits the commands over the CAN Bus to the Motor ECU's MO1_ECU/1 and MO2_ECU/1, respectively. The Motor ECU's generate the control variables for their attached motors, according to the driving commands received from the CMD_ECU and the actual motor values that are fed back over the CAN network via the super-elements RO1_ECU and RO2_ECU, respectively, attached to the motor outputs.

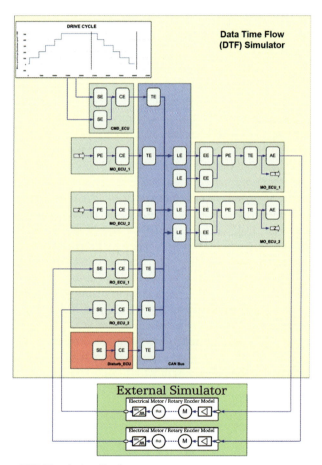

Fig.4.   DTF Simulation Environment

In Figure 4, the CMD_ECU, two super-elements MO1_ECU/2 and MO2_ECU/2 can be seen. These super-elements logically belong to their counterparts MO1_ECU/1 and MO2_ECU/1. Their only task is to send CAN frames received from their attached Motor ECU's over the CAN network. These frames contain the current nominal and actual rotation speed for each motor as well as the current motor control variable. The purpose of these frames is to generate traffic and to allow for observation of the current state of motor control over the CAN network (e.g. via a CAN frame analyser attached to the network like in the CDP).

Finally, we provide a disturbance node Disturb_ECU (left-hand side at the bottom) that allows for intentional disturbances of the CAN network. The disturbance node can be used to expose the CAN network to stress by generating additional traffic. Further, a disturbance node may exhibit various failure modes like e.g. a babbling idiot failure.

3.3 Evaluation

For a first validation experiment, we defined a simple driving cycle. The driving platform accelerates to 50% of the maximum speed, keeps the speed for a while and then gradually decelerates to idle speed.

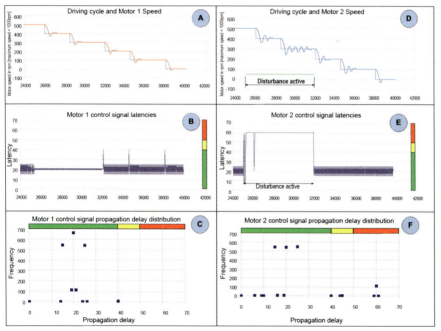

Fig. 5.   DTF Simulator result, with/without disturbances

During the deceleration phase a network overload, performed by the disturbance node, initiates a frame delay which degrades the closed-loop control performance of the motor 2 speed control. The related simulation results are shown in Figure 5. The diagrams on the left side shown a motor control sequence without influences by the disturbance node (Diagrams A, B and

C), while the diagrams on the right side ((Diagrams D, E and F) shows the behaviour with extended delays of the CAN frames during transmission on the network. In diagram D the resulting swing in the actual speed is clearly identifiable. Diagram E shows the transmission latency time over the full simulation time and diagram F shows the latency time distribution during the full simulation drive cycle. In both diagrams the specified green areas are exceeded and the yellow and red areas are met. The correct function of the closed-loop control can not be guaranteed.

This behaviour was widly identical with the real values, measured on the real hardware of the Concept Development Platform CDP.

To sum up, this simulation, performed in a very early development phase, e.g. in the network concept design phase, will give very important hints to find the optimal network architecture. Several concepts and architectures can be evaluated and assessed without the existence of detailed design data. The best fit concept network designs are reviewed in a concept design assessment process. In the subsequent design steps the modelling of the network architecture get more and more in detail and an continuous design validation against the requirements would become reality.

References

[1]   Wagner, F., et al., Simulation of Distributed Automation Systems in Modelica, Modelica Conference, University of Kaiserslautern, 2008.

[2]   Lisner, J., Kessler, P., Entwurf eines TTP/C-Feldbusmodells mit der Spezifikationssprache SDL, Technischer Bericht Nr. 5, Universität Essen, Institut für Informatik, 2001.

[3]   Echtle, K., et al., Konformitätsbewertung für sicherheitskritische Anwendungen im Umfeld fehlertoleranter Datenbus-Systeme; Stand des Forschungsprojektes „FlexBeam".

[4]   Millinger, D., DECOMSYS; Model-Based Design of Automotive RT Applications; Präsentation Embedded Systems 2003.

[5]   MATLAB Select, X-by-Wire Systementwicklung in Simulink, Vol. 1, 2003.

[6]   Architectures for Safety; dSPACE News, 1, 2007.

Alexander Hanzlik
Simulation Competence Center
Apostelgasse 39
1030 Vienna
Austria
ahanzlik@gmx.at

Erwin Kristen
AIT Austrian Institute of Techology GmbH
Donau-City-Straße 1
1220 Vienna
Austria
erwin.kristen@ait.ac.at

Keywords: simulation, automotive control networks, CAN, FlexRay, data time flow, safety, concept design assessment

Dynamic Cruising Range Prediction for Electric Vehicles

P. Conradi, P. Bouteiller, Steinbeis Innovation Centre E-Mobility
S. Hanßen, ALL4IP Technologies GmbH & Co. KG

Abstract

Battery electric vehicles (BEVs) require new driver information systems. We anticipate a new integrated and networked information system class, combining data input from central car systems, drivers' behaviour and environmental parameters. By introducing the system mapZero we propose an OEM-independent cruising range prediction system, which combines measurement and GIS-system based calculations on the ride (see Fig. 1). For the first time, we can consider range-affecting variables like outside temperatures, driving style preferences, charge metering, navigation, consumption planning, and route distance prediction to manage the immanent range restrictions.

The car side system consists of an optional on-board unit with metering and communication properties, and a mobile unit, preferably a smartphone (e.g., iOS, Android), offering navigation, power management, information and convenience features. System core is a self-learning algorithm as aggregation feature inside of a collaborating community, the car's actual performance behaviour and the users preferred destinations and driving styles. It will propose charging or battery swapping stations on the planned route, which will then be passed on to a reservation management system, making sure the driver can enjoy a maximum of comfort and peace-of-mind on his or her journey.

1   Introduction

In times of limited oil resources, a swift progression to carbon-independent mobility is more important than ever. The general assumption is a sequential introduction of new technologies depending on car segments, i.e., we will see a rapid replacement of small and medium range cars in city areas and commercial fleets with pure electric drives, whereas long range cars will use some form of more expensive and complex hybrid technologies.

Range optimization will become crucial to support our mobility requirements. New forms of mobility will become more acceptable for large parts of the population and we will see a more widespread introduction of co-modality systems (integrated combination of private and public transport) as well as car sharing concepts [2]. Car sharing and car rental systems will be particularly dependent upon intelligent information systems. For instance, during use the predicted capacity of the battery will be transmitted to a central server in order to manage recharging intervals to provide sufficient capacity for the next user. These are some of the challenges that require central computing capacity and smart real-time information systems for fleet management, logistics and the general driving experience.

Fig. 1.   End User App in Galaxy Pad

Driver information systems for range prediction will be even more important as it currently looks unlikely to expect a nationwide charging infrastructure soon, based on the currently preferred technological solution: metering and measuring technology inside charging stations makes them prohibitively expensive and commercially unviable. Moreover, electric cars will come with a wide range of individual differences: There will be many models, types, batteries, charging states, ages, etc. The unique state of each individual system will make smart driver information systems so much more important. Perhaps this would also be a good time to think about community based information systems and tap into the power of the wisdom of the crowd.

2 Cruising Range Prediction

Range is a highly sensitive topic when it comes to electric vehicles. Besides the cost, the limited driving radius is a significant barrier to the spread of electric vehicles. High prices, a general reluctance of the industry, prejudices and stereotypes also influence the development of electric vehicles since their inception. What is to be done?

First, the buyer has a natural desire to know what he gets for his investment. Since the range of electric vehicles varies greatly according to use, it is important to provide information systems that are associated with the user and mobile. These systems connect to the vehicle and deliver relevant information to the user about core functions of the vehicle.

These include firstly:
- ▶ Reliable range forecasts
- ▶ Management of the recharging process with reservation, vehicle dock arrangement and energy billing

Based on this basic concept, various additional services can be added:
- ▶ Tools for the organisational design of daily routines of the user, based on the energy state of the system
- ▶ Mobility planning with alternative modes of transport, including car sharing or public transport (Intermodalität [5])
- ▶ Recommendations for environmentally sound customer-specific charging strategies

In this article, the authors focus on the range analysis in electric mobility. As the X-chart in Fig. 2 shows, a number of variables determine the potential reach of an electric vehicle with its energy supply. In addition to the design factors of the vehicle, the users' driving patterns are crucial, but also environmental conditions play a significant role in the range estimation.

Thus, the authors propose a system for the range prediction analysis, which uses energy storage and energy release in the vehicle, the vehicle dynamic data, driver behaviour, environmental conditions and route specifics.

Environmental computations take place based on a server-based geo-information system that uses both topographical and topological data. The topology server will compute energy calculations based on the topological node / edge-model of the route. Wind components and height differences can be considered for the computation of the range map model (a graphical representation of the predicted range), as well as typical speeds and speed limits, all depending on user selections in the GUI.

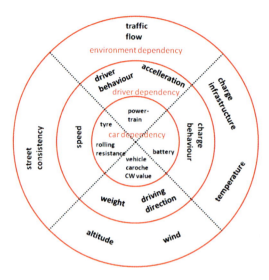

Fig. 2.   X-Chart of E-Mobility

A topographical search will start by identifying the next crossroad from the current position. Then a topological search-algorithm will consider all possible roads leading away from that position to the respective edges of the range map and computes the energy consumption based on GPS-determined driving dynamics. In this calculation, every route is calculated under consideration of the car specific energy model and subtracted from the base energy. Thus, one derives a range prediction graph as a polygon, which will constantly shrink as one drives along. Depending on the car's velocity and charging state (SOC), the polygon will be mapped in colour onto the street map with adjusting accuracy and refresh rate.

The total system consists of a series of components on a cloud server that are configured via four human access interfaces, (1) to (4), shown in the architectural scheme in Fig. 3.

(1) The Web Range Estimator addresses the energy consumption calculation based on defined standard profiles. Independent of position, and thus independent from the topology server, it will deliver a range prediction considering the vehicle load, temperature and battery age.

(2) The App (Program on mobile device) runs inside the vehicle with GPS-localisation and GSM communication and dynamically displays the position of the vehicle and the map as received from the environment server, as shown in Fig. 1. The red colour shows the reachable area. In the lower right side, vehicle direction and wind is displayed in a compass-similar arrangement. The App communicates via TCP/IP in order to deliver vehicles state and position to the topology server, and receives and displays the dynamically computed range polygon on the map.

Web Range Estimator and Topology Server both make use of the energy calculator for their range computations, in order to derive the energy consumption prediction for each route.

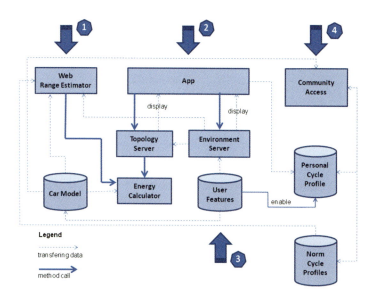

Fig. 3.   mapZero Architecture

(3) Via a User Feature-database each user can integrate value added services (like additional calculation options or graphical display features), which can be purchased as add-ons.

(4) The Community Access will allow every user to rank publicly released profiles and compare them to standard profiles. This could lead to community-acknowledged standard profiles for different vehicle types, independent of industry data.

3   Car Model Data Base

The car model database currently contains over 30 car model descriptions. By using the simulation mode the cars can "drive" through different standard driving cycles, comparing the respective velocity / time tables. Parameters such as age of battery, weight, and temperature allow for calculating the range based on these standard driving cycle patterns EUDC [5], CADC [3], and TSECC [1].

4   App and Polygon Server

The range information that is visualised on a map is the clear focus of the creators and was developed for both, two-wheeled and four-wheeled electric vehicles. It allows topological information about the technically possible and the natural limits to individual E-Mobility. Just imagine that the app not only depicts charging stations or the next bakery, but also the specific limits of the respective electric vehicle.

5   Community Solution

As the article [2] (with a commentary by the author at the end) clearly shows, the vehicle manufacturers information transparency should meet the requirements of the slowly developing customer base of electric vehicles. It should lead to a process of awareness for the customer regarding available community knowledge, which would have more credibility and draw more attention to the product than standard advertising measures by the manufacturers. Moreover, it would allow evaluating technical qualifiers (such as the driving range) much better than the manufacturer.

This effectively means that members of the community can evaluate available electric vehicles in correlation to their respective travel profiles based on crowd-based information. Any member who has an electric vehicle will contribute to optimizing the prediction database. Thus, the whole community contributes to testing and improving the quality of the system's prediction quality. In turn, each published profile will be available to all other users free of charge. Particularly in areas where standard profiles fail this will be a welcomed benefit for the community. Over time, a crowd based community system will develop, which would be far superior to any other information source. Today, over 30 vehicles can already be tested with the system.

6   Value Added Services for the End Customer

6.1   Basic Application

In basic mode the user computes the range, resulted by the Web Range Estimator. Using this length value the app estimates the range by putting the polygon together with street parts in the driver's environment and adjusting the state of charge in a linear way. To do so, the app calls the topology server and shows the polygon on the smartphone display.

6.2   Advanced Application, Using the Car Model

The range display leads to more precise results, if the respective car model is computed. This service can be ordered as a user feature. In this mode the range is computed directly on base of the technical data of the vehicle. This is sufficient if the operational data of the vehicle are known and the vehicle is in a defined charge state, in the ideal case fully charged. In this case the topographical range can be computed without access to the vehicles electronics.

6.3   Application with Physical Adaption

The next step of precision can be reached by installing a data bridge to the board electronic. By applying a wireless connection to the CAN-bus, the State of Charge is transmitted to the smartphone. This can be achieved for all vehicles with CAN bus and known ID. Since this is the only way to differentiate between (partial parallel) drive, the physical measurement is recommendable especially for hybrid vehicles, e.g. pedelecs. Not all vehicles of a series need a measurement module. It is sufficient if only a few vehicles are equipped with an adapter and the electric vehicle experience the typical behaviour. These experiences could be automatically used by the other community members with the same type of vehicle to obtain high quality predictions. Via the right mix of resources and distributed measuring the reliability of the overall system can be increased.

The regularly updated range polygons on the smart phone display allow for an immediate and intuitive assessment of the energy situation and a smart planning of reachable destinations, charging breaks, or a change of the mode of transport by jumping on a bus or a train [5]. The "point of no return" is always in the visual view of the driver. Also, he or she will always know, where the next available charging station is located and will, thus, have a comfortable and relaxed driving experience.

7   Conclusion

The increasing precision in range estimation can lead electronic mobility into a new age. Using crowd-based intelligence enables the amalgamation of difficult to collect data into reliable predictive values. This allows a much more reliable planning of journeys and counters concerns around battery depletion and the corresponding break down of electric vehicles. The fact that drivers are using smart phones which they carry around at all times, leads to new degrees of freedom and will increase the acceptance of electronic mobility in the public.

8   References

[1]   Bloch, A., Eiszapfen (in German), Auto Motor Sport, p 142 - 147, Motor Presse Stuttgart, Dec 16, 2010.

[2]   Bouteiller, P., Conradi, P.: Advanced Mobile Information and Planning Support for EVs (MIPEV). In: Meyer, G., Valldorf, J. Advanced Microsystems for Automotive Applications 2010, Smart Systems for Green Cars and Safe Mobility, Springer, Berlin, 2010.

[3]   Common Artemis (Assessment and Reliability of Transport Emission Models and Inventory Systems) Driving Cycle, Result of an EU Project, published in Report INRETS-LTE 0411, June 2004.

[4]   EEC Directive 90/C81/01

[5]   Maertins, C. , Die intermodalen Dienste der Bahn (in German), Discussion Paper SP III 2006-101, Wissenschaftszentrum Berlin für Sozialforschung, 2006.

[6]   Noveck, B. S.: Wiki Government: How Technology Can Make Government Better, Democracy Stronger, and Citizens More Powerful. Washington, D.C.: Brookings Institution Press, 2009.

[7]   Surowiecki, J.: The wisdom of crowds , 2004.

Peter Conradi
Steinbeis Innovation Centre E-Mobility
Lise-Meitner-Str. 10
64293 Darmstadt
Germany
peter.conradi@stw.de

Philipp Bouteiller
Leibnizstr. 55
10629 Berlin
Germany
philipp1@mac.com

Sascha Hanßen
ALL4IP Technologies GmbH & Co. KG
Robert-Bosch-Str. 7
64293 Darmstadt
Germany
sascha.hanssen@all4ip.de

Keywords: mapZero, E-Mobility, electrical vehicles, BEV, range, reach, SoC, state of charge, driving range, range prediction, app, cloud, topological analysis, GIS system, map, polygon, X-Chart.

Components & Systems

iGMR-Based Angular Sensor for Rotor Position Detection Within EC-Motors

M. Weinberger, W. Granig, Infineon Technologies Austria AG

Abstract

Due to the hybridization and change of hydraulic systems into electrical systems, the number of electric commutated motors in vehicles is increasing dramatically. There is a trend towards brushless motors, as they have no wearing parts (brushes), low acoustic buzz and very low torque rippling. In order to ensure exact commutation of electric motors the position of the rotor has to be measured very accurately over a wide speed range. The position is afterwards transferred to the control unit, which generates the necessary commutation sequence for the motor. In the last years the position sensing was done by Hall switches for block commutation. Nowadays, the call for absolute position sensors for sinusoidal commutation in motors is getting louder. Giant Magneto Resistance (GMR) angle sensors [1-4] offer a perfect solution.

1   Introduction

Integrated GMR (iGMR) angle sensor can be used for various motor topologies with different numbers of pole pairs. Instead of three Hall switches that have to be positioned very well only one sensor has to be arranged. This saves wires and costs. Figure 1 depicts the easy setup with the iGMR angle sensor and a diametrically magnetized magnet.

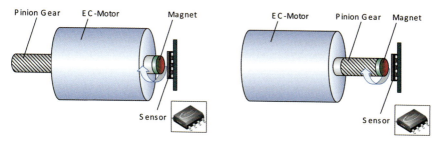

Fig. 1.   Two possibilities for placing magnet to sensor

This type of rotor position determination at various actuator drives in the automotive sector has already been successfully implemented.

2   iGMR Angular Sensors

Infineon's angle sensors are based on the Nobel prize awarded GMR effect. This technology can be integrated into standard automotive qualified semiconductor products. The GMR structures are vertically arranged above the logic portion [1], hence the designation of integrated GMR sensor (iGMR).

Four individual GMR elements are connected as a Wheatstone bridge and measure one component (e.g. sine component) of the external magnetic field. By implementing another Wheatstone bridge the second component (e.g. cosine component) can be measured. With the use of the trigonometric arctangent function "ATAN", the absolute angle can be determined.

This type of angle sensor offers calibrated angle signals on the output. It is optimized for brushless DC motor drives equipped with many features for safety critical applications. This sensor includes the monolitically integrated GMR bridges, of which signals are converted into the digital domain by Sigma Delta converters (ADC SD) immediately. Subsequently the signals are corrected by calibration data and the absolute angle is calculated. This angle information can be read out via different interfaces.

2.1   Autocalibration

For an efficient commutation the "electrical" angle error must be at least smaller than 10° (application-dependent). The required mechanical accuracy however depends on the structure of the motor. Depending on the number of pole pairs the mechanical angle error has to be many times smaller than the electrical one. To achieve this high mechanical accuracy, a special algorithm was implemented, which adjusts the sensor calibration parameters during operation automatically. This algorithm is a so-called 'auto-calibration'.

Using the auto-calibration, it is possible to reach an accuracy of smaller than 1.0° over temperature-variations and operating lifetime. This auto-calibration continuously calculates the latest parameters such as amplitude and offset during operation and calculates the new absolute angle using them. Figure 2 shows the angle error over temperature and different external magnetic fields.

By activating the auto-calibration (enabled or disabled via Synchronous Serial (SSC) Interface) the angle error can be reduced significantly.

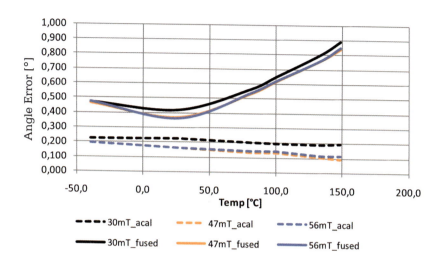

Fig. 2.   Auto-calibration vs. fixed parameters

2.2 Signal Delay

Furthermore it is important to get the current angle information as soon as possible. During the detection of the position by the GMR sensors, calculating the absolute angle, transmission of the angle and further processing in the electrical control unit (ECU), the motor continues to rotate. This results in an additional 'dynamic' angle error. To keep this angle error as small as possible each delay has to be reduced as far as possible.

Update-Rate: The update rate of the sensor can be selected between 42μs, 85μs and 170μs. High update rate signals are faster but more noisy, smaller update-rates are slower but have less noise.

Prediction: To minimize the additional dynamic angle error, an additional feature can be used, the so-called 'prediction'- function. This is a linear prediction of the angle (Equation 1).

$$\alpha(t+1) = 2 \cdot \alpha(t) - \alpha(t-1)$$

(1)

A reduction of the delay time can be achieved and so a minimizing of the additional dynamic angle error is possible [5]. Figure 3 depicts this function for a better understanding. In this figure a change in direction is shown. Without prediction the sensor will follow the mechanical rotation slowly. The advantage of the prediction can be seen easily. With activated prediction the sensor will follow the external magnetic field much better than without prediction. In this figure an update rate of 42 μs was assumed.

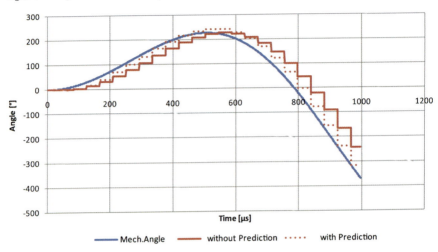

Fig. 3.   Output signal w/o prediction

Synchronization: By using an external clock input signal the sensor can be operated completely synchronous with the microcontroller. The advantage of this is to select the timing for the read out command. So the latest angle information with the lowest delay time can be read out (Figure 4). This can be verified online by reading the Frame-Sync continuously.

2.3   Interfaces

The angle value can be read out through various interfaces. Via Synchronous Serial Interface (SSC) the sensor can be configured and also various information can be read out. This interface is compatible with standard serial peripheral interface (SPI), the only difference is the data line, which can be used bidirectional. The additional open-drain functionality allows to setup a sensor bus system operating with up to four sensors.

Beside the most important parameter angle, additional information is available like angle speed, number of revolutions and status information. This information could be important for motors in safety relevant applications. To further reduce delays, the SPI transfer can be operated with 8Mbit/s via a Push-Pull configuration.

3 Hall switches are frequently used in brushless DC motors to roughly detect the rotor position. Depending on the pole pairs the switches must be arranged in a certain geometric distance from each other. Through this arrangement, a typical output signal pattern is generated which is shown in Figure 5. These signals can be processed very simply for block-commutated motors.

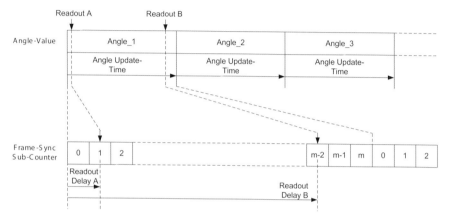

Fig. 4.   Synchronous read out to reduce delay

The GMR angle sensor has the possibility to emulate these 3 output signals. Thus it is possible to save 2 additional sensors and the exact arrangement of the three Hall switches to each other. The amount of pole pairs can be easily configured via SPI access. Motors with up to 16 pole pairs are supported. This emulation of Hall switches is called as Hall Switch Mode (HSM) and can be used in parallel to the SPI interface. Figure 5 shows the output stage. Depending on the pole pairs of the motor, the switching in accordance to the mechanical position occurs sooner or later.

This interface is very well suited for high speed. Also an advantage at high speed is the Incremental Interface (IIF). This one has the possibility to determine the absolute position by counting pulses on ECU side. Two different modes are available via SPI: A / B-Mode (optical encoder standard) or Step / Direction Mode (Figure 6).

The A / B mode puts pulses on two tracks. These pulses are phase shifted by 90° electrically. The controller uses these pulses as counter input. Depending on the direction of rotation Phase A follows Phase B or converse. This is recognized in the counter and the count gets increased or decreased accordingly.

In Step / Direction Mode the direction is not determined by order of the pulses. A separate direction-track indicates the counting-direction. The other one is used only for counting, depending on direction.

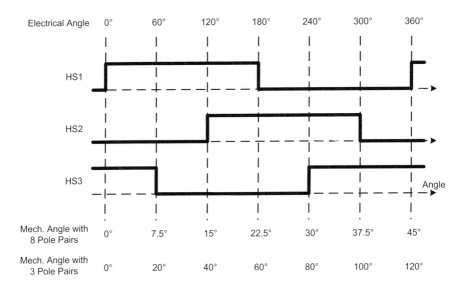

Fig. 5.   Hall Switch Mode: 3phases Generation

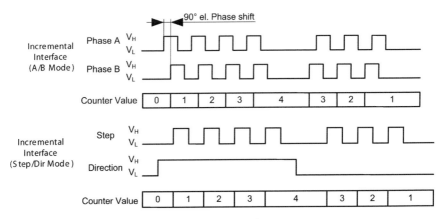

Fig. 6.   Output signals of incremental interface

2.3 Performance

In Figure 7 the performance of the GMR angle sensor is presented. The sensor was exposed to a homogeneous magnetic field of 30 mT at 25° C. To determine the hysteresis, the sensor was measured in a clockwise direction and also counterclockwise. The difference between the two curves is the hysteresis of the sensor. This sample results in an angular error of about ±0.4° over the entire angular range of 360°. The same measurement was carried with this sample again at 150° C and can be seen in Figure 8. In both measurements the angle calculations were done with the calibration parameters at room temperature.

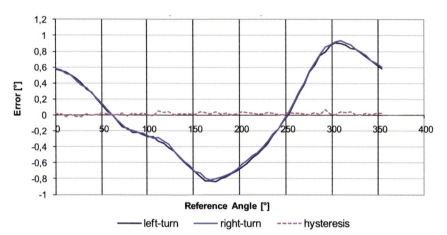

Fig. 7.   Output signals of incremental interface: Angle Error Distribution (@150° C, 30 mT)

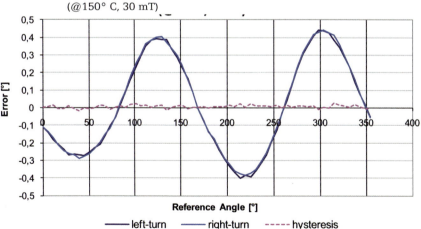

Fig. 8.   Output signals of incremental interface: Angle Error Distribution (@25° C, 30 mT)

3 Summary

Angle sensors are a good solution to relieve the controller from time-consuming computations. Also for safety relevant applications they offer various safety features.

For motors with high speed the immediate availability of the angular values is essential. The high update rates and additional useful functions as well as the prediction and synchronization minimize the delays and reduce the additional angle error.

iGMR angle sensors can be used for various motor topologies with different numbers of pole pairs. Instead of three Hall switches, which have to be positioned very well to each other, only one sensor with IISM is necessary. Number of used wires, time and costs can be minimized. The iGMR angle sensors offer therefore a very good alternative to Hall switches.

Because of the implemented IIF also optical encoders can be replaced inexpensively.

Angular error caused by the sensor element can be kept low by a precalibration. Another angle error reduction is achieved by a production-calibration in the target application. A special feature in order to achieve the lowest possible angular errors in operation and over temperature-variations is the so called auto-calibration. Additional errors can be caused by the magnet system, which must also be considered precisely.

References

[1] Hammerschmidt, D., Katzmaier, E., et al., Giant magneto resistors – sensor-technology and automotive application, SAE2005 Detroit USA Nr. 05AE-127, 2005.

[2] Granig, W., Kolle, C., et al., Integrated Gigant Magnetic Resistance based Angle Sensor, IEEE-SENSORS 2006 Daegu KOREA, Nr. B1L-C4, 2006.

[3] Granig, W., Hammerschmidt, D., Anastasiadis, I., Magnetischer Splitter für Magnetoresistive Winkelsensoren, www.ip.com, 10/2007.

[4] Granig, W., Köppl, B., Hartmann, S., Performance and Technology-comparison of GMR versus commonly used Angle Sensor Principles for Automotive Applications, SAE2007 Detroit USA Nr. 07AE-151, 2007.

[5] Hammerschmidt, D., Granig, W., Salzmann, S., Predictive Angular Sensor Readout, US2009/0326859A1, 2009.

Markus Weinberger, Wolfgang Granig

Infineon Technologies Austria AG

Siemensstraße 2

9500 Villach

Austria

markus.weinberger@infineon.com

wolfgang.granig@infineon.com

Keywords: iGMR, TLE5012, angle sensor, rotor position, EC-motors, prediction, auto-calibration

Architecture of an Integrated AMR Current Sensor (IACS) System for a Wide Range of Automotive Applications

A. Nebeling, W. Schreiber-Prillwitz, Elmos Semiconductor AG

Abstract

This article describes a new anisotropic-magneto-resistive-effect (AMR) based current sensor system, based on a 'system-in-a-package' approach. It is designed for high resolution and fast electronic measurement of DC, AC up to 500 kHz, or pulsed currents [1]. As the concept is based on an external primary current bar, it can easily be adapted to a wide range of applications for different nominal primary current ranges by just altering the geometry of the primary current bar. The major advantages of this system are its high bandwidth and high dynamic range offered by the AMR measurement principle combined with a negligible hysteresis and improved linearity features. The possibility of end-of-line calibration in the application environment at the production site allows for an optimum sensitivity adjustment and improved system accuracy. The integration of the current sensing components within one standard plastic package allows the cost efficient high volume assembly for the customer.

1   Introduction and General Description

As an alternative or a replacement for Hall sensors, AMR based sensors are increasingly entering the field of angular and linear position measurement, as well as current measurement. For current measurement, Hall effect based sensors normally are utilizing an iron core to concentrate the magnetic field lines to the sensor area. This flux concentrator typically encloses the primary current line, with the Hall element placed in the air gap of the flux concentrator. This construction leads to a perpendicular orientation of the concentrated magnetic field lines to the sensor surface, and by this method to a sufficient signal. Hall based systems without flux concentrator do need a primary current line inside their housing to enable the now needed small distance between Hall element and current line to achieve a sufficient signal. The necessary small distance on one hand decreases the dielectric strength of those products, on the other hand the fix cross section of the integrated primary current line determines the maximal allowed current. The bandwidth range of Hall effect based current sensors is typically 50 to 150 kHz. In general, to cover different ranges of current, different products are needed.

The described AMR-based current sensor exhibits no hysteresis as observed in iron core based Hall sensor solutions. Due to the high sensitivity of AMR sensors, a flux concentrator is not necessary [2]. It is designed for high resolution and very fast electronic measurement from DC up to 500 kHz AC. Contrary to Hall effect based sensors, the described system enables a differential magnetic field measurement by an advanced geometry of the magneto resistive elements. Due to this construction the sensor is immune to homogeneous interference fields [3].

By variation of the geometry of the external primary current bar, this system can be adapted to different current ranges and applications. With this, only one system is needed for a wide range of applications.

The system accuracy can be improved by using either the internal or an external reference voltage. This further reduces temperature drift, and several sensors can share the same reference voltage. The adjustable over-current detection enables fast response in overload situations to prevent damage to e.g. power units.

2   Demands of Future Applications

The development e.g. in the field of regenerative energy generation will bring a lot of new application fields for current measurement. Power inverters for photovoltaic panels, or current monitoring in the field of electro mobility (motor control, battery charging and monitoring) are only the most obvious ones. Fact is, that future applications will need to monitor currents in the range from Amperes up to several hundred Amperes, at high bandwidth and simultaneous high dielectric strength. This demands a product with high accuracy and inherent high flexibility.

3   IACS System Concept

The new sensor system developed by Sensitec and Elmos, based on the anisotropic magnetoresistive effect (AMR-effect), fulfills these demands. It consists of the AMR sensitive sensor cell, the signal conditioning circuitry and two biasing permanent magnets. The latter are for maintaining the initial magnetization direction of the AMR structures after the application of very strong over-current spikes. The permanent magnet raw material and the AMR-sensitive sensor material are applied on wafer substrates by a special process, thus can be processed further on with standard semiconductor methods, concerning singulation or assembly. A special leadframe as well as an advanced assembly technique enables a 'system–in-package' solution (SIP): all system components are overmolded within a JEDEC compliant SOIC16 package (Fig. 1). The product can be mounted with standard pick-and-place equipment onto a PCB, and subsequently reflow soldered (Fig. 2).

Fig. 1.   IACS system concept

Fig. 2.   Reflow soldered IACS

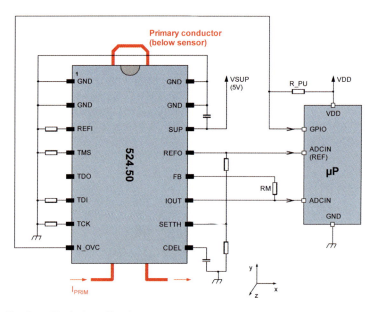

Fig. 3.   Typical application

The magnetic field is changing the resistivity of the elements of the Wheatstone bridge. This imbalance of the bridge results in a differential voltage which will be amplified by the read-out IC and converted to a proportional current at the output of the IC. The IC is feeding this proportional current into the compensation current line of the sensor cell, which will create a magnetic field of the same magnitude, but opposite direction as the field generated by the primary current line at the sensor position. A block diagram of the read out IC is given in Fig. 4. With this closed loop principle, high linearity and zero temperature dependency of sensitivity is achieved, as the sensor always is working close to zero excitation. This contributes significantly to the high accuracy of the system. The use of fast, low noise amplifiers allows short step response times <1 µs. The output noise of <1.5 µA eff. within a bandwidth of 200 kHz results in an SNR of >66 dB. An overload of 3-times the nominal current is allowed for short pulse times (<50 ms), in which the system still delivers precise output. Together with the high bandwidth, not only average values, but also fast current transients can be measured. This opens the field for time critical applications like e.g. current sensing in IGBT-bridges used in power inverters.

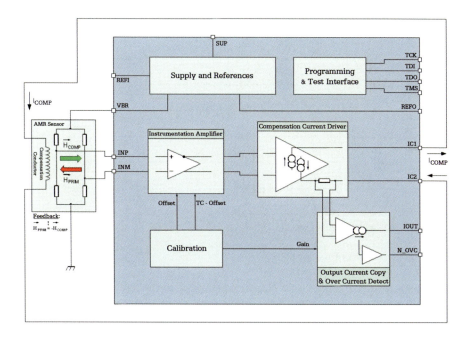

Fig. 4.   Block diagram of the signal conditioning circuitry

4   Technical Base of the IACS

The AMR effect describes the variation of the specific resistivity of a material, depending on the angle between an electrical current through this material and an external magnetic field. If the external magnetic field and the current through the material are parallel (0°), the specific resistivity reaches its maximum, in case of perpendicular orientation (90°), $R = R_{min}$. The variation of the resistivity is in the range of 1 to 3% and has quadratic characteristics (Fig. 5, red curve). A rotation of the current through the AMR-sensitive resistors of 45° towards the length axis leads to a more linear transfer function around 0° (Fig. 5, green curve), and gives additionally information about the direction of the field. This rotation can be achieved by so called 'barber poles'. These are realized as metal lines (Al) under 45° upon the length axis of the permalloy material.

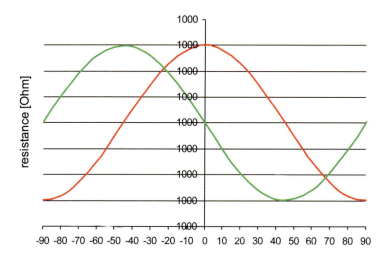

Fig. 5.   Effect of barber poles

The current sensor is built as a Wheatstone bridge with a certain geometric set up: both half bridges are located in areas with equal distance from a symmetry axis of the sensor (Fig. 6). A different barber pole orientation within each branch (R1 + R2, R3 + R4) leads to a contrary variation of the resistivity within the branch, and so effectively raises the sensitivity.

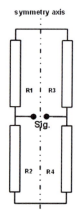

Fig. 6.   Wheatstone bridge configuration

The AMR sensitive structure of the IACS consists of a permalloy, which has a natural form anisotropy concerning its magnetization due to its dimensions (some 10 nm thin, some µm wide, several 10 µm long). The magnetic domains are oriented along the longitudinal axis of the structure. The effective magnetization results from this natural internal magnetization and the applied external magnetic field. As the internal magnetization has no preferred direction along the longitudinal axis, a flipping of 180° could occur due to overcurrent spikes. This would lead to a different effective magnetic field, and so to a different characteristic of the system compared to its original configuration. In order to always orient the internal magnetization to a defined direction, the biasing magnets are needed.

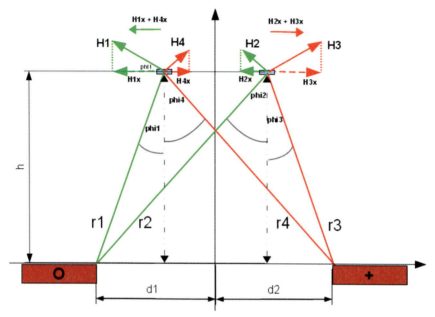

Fig. 7. Effective Hx-field at the two half bridges of the sensor

4.1 Gradient Measurement

The design and the geometric placement of the primary current bar have a big influence on the field gradient over the sensitive area. The recommended U-shaped conductor contributes with four H-field vectors to the resulting magnetic field at the sensor, as is illustrated in Fig. 7. The light blue upper bars represent each one branch of the bridge (see also Fig. 6). The brown bars show the cross section of the primary current bar with a distance d1 and d2 to the symmetry axis of the sensor. The orientation of d1 and d2 represent the x- axis in the following diagrams. If d1 and d2 are identically, the symmetry axis of sensor and current bar are aligned, and the geometrically offset is zero. In this simplified sketch it can be seen, that each part of the sensor sees a different resulting Hx-component of the magnetic field generated by the primary current bar. The sum of both absolute values represents the magnetic field gradient over the distance of the AMR-sensitive areas of the Wheatstone bridge configuration.

4.2 Immunity against Stray Fields

One big advantage of measuring the field difference between the two positions of the half bridges is the immunity against external stray fields.

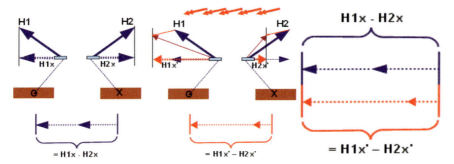

Fig 8.   Effect of Magnetic Stray Fields: a) resulting undisturbed gradient Hx (left), b) resulting disturbed gradient Hx (middle), c) resulting grad Hx (right)

Fig 8a) shows the resulting field gradient Hx in the undisturbed case. The effect of a stray field is shown in Fig 8b). Its components (red arrows) are adding the same magnitude to the Hx vector component for each side of the sensor, and the resulting magnetic field gradient keeps the same (Fig 8c) under the assumption, that the stray field is homogenous over the area of the sensor chip. This assumption is plausible due to the small distance between the resistors in the 1 mm-range.

5   Realizing a Wide Range of Applications

The concept of an external primary current bar gives a certain degree of freedom to adapt the system to a wide range of applications. To get best SNR, the field gradient should be near 1600 (A/m)/mm for the full scale range of the primary current in a symmetrical arrangement of primary current bar and sensor area. A second aspect becomes obvious from Fig. 7. The lateral expansion (width and thickness) of the primary current bar can not be neglected, as the different parts of the cross section will contribute different to the resulting magnetic field at the sensor. So in reality, it is not sufficient to only calculate the magnetic field after the Biot-Savart formula for a thin linear wire, but to integrate over the cross section of the primary current line. This is even more important for current lines for high currents (several hundred Amperes), whose cross sections will have dimensions in the range of mm, and so are in the same range as the sensor geometry itself. The characteristics of the gradient within the xy-plane should be constant within an area of typical lateral assembly deviations, as this would lead to only small variation in sensitivity. Fig. 9 shows the trend of the magnetic field gradient in (A/m)/mm, when the sensor would be moved along the x-axis. This represents the influence of lateral displacement between sensor and primary current bar (with $x=0$ as the symmetric target configuration in Fig. 6). Additionally, in Fig. 9 the cross section of the primary current bar is changed (see legend). The green line would be the chosen geometry, as it has a low variation to symmetry displacement around $x=0$, and comes closest to the target value of 1600 (A/m)/mm. A certain variation of the gradient value due to assembly related vertical distance variations can not be avoided. The possibility of an end-of-line calibration of the gain by EEPROM programming after assembly compensates this effect.

Trend of grad Hx vs. x-displacement

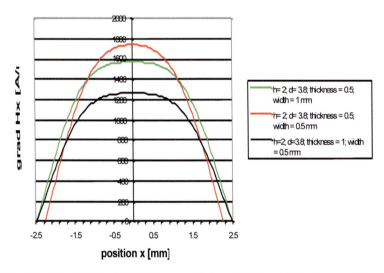

Fig. 9.   Trend of Grad Hx for Inom = 50 A and Different Prim. Current Line Cross Sections

In Fig. 9 it can bee seen that the thickness of the primary current line is a sensitive parameter concerning the value of the field gradient (red: thickness = 0.5 mm, green: thickness = 0.5 mm, black: thickness = 1 mm). In general, the design of a primary current line has to take into account the current load a certain cross section of a conductor can carry, especially concerning temperature rise. The IPC-2221 for instance gives guidelines how to calculate the cross sections for copper current lines embedded into PCB or attached onto the PCB surface (I < 35 A, trace width < 10 mm; beyond these limits the formular extrapolates the values) [4].

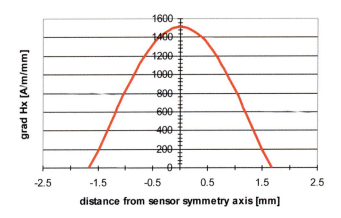

Fig. 10   Trend of Grad Hx for Inom = 25 A

Table 1 shows the calculated minimum trace width after this formula for a nominal current of 25 A for an external copper current line. Fig. 10 gives the according grad Hx trend for this geometry. From this it is possible to realize a current measurement system for Inom ranges from 0 A to 25 A up to ranges above 0 A to 200 A, by having the full signal resolution.

current	25	A
Cu-thickness	0.4	mm
T- increase	20	°C
min. trace width	1.5	mm
vertical distance	1.5	mm
distance between arms	1.2	mm
max. grad H_x	1500	A/m/mm

Tab. 1.   Dimensions and geometry for Inom = 25 A

5.1 Preliminary Target Performance

Parameter	IACS 524.50
Nominal input (peak input)	+/- 1.25mT (+/- 3.75mT)
Nom. Output (Peak output)	+/- 2mA (+/- 6mA)
Sensitivity	1.6 mA/mT
Bandwidth / step response time	500kHz / <1µs
Output noise (@BW)	1.5 µA $_{rms}$ (BW = 1Hz ... 200kHz)
Sig.Nom / Nrms	1333
Offset in Temp (% Nominal)	+/- 0.75%
Sensitivity drift (in Temp)	< +/- 1.0%
Linearity error (% Nominal)	< +/- 0.3% @ OUTn < +/- 1.5% @ OUTpk *

* The linearity error refers to the nominal output of 2 mA, also for the peak output of 6 mA.

Tab. 2   Prel. target performance key parameters for IACS

The SNR of 63 db is the base of the parameter sig. Nom/Nrms = 1333. With this, a magnetic field gradient of 1.2 (A/m)/mm can be resolved, if the target value of 1600 (A/m)/mm is reached at Inom. A current in the range of 100 mA could be resolved for the above example in Fig. 10.

5.2 First Results

The following figures show first measurements of engineering samples with all system components enclosed within one SOIC16 package. In Fig. 11 the high dynamic of the system is demonstrated. First measurements of the bandwidth in Fig. 12 show that the targeted bandwidth of 500 kHz is exceeded.

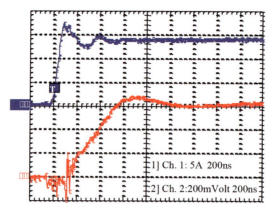

Fig. 11   Signal response. Rise time <600ns

6   Application Example

Fig. 13 shows the use of the IACS for a three phase rectifier for high currents (~200 A). This example shows the benefit of the external current bar concept: the PCB is fabricated with standard methods, with also the current sensors attached by standard reflow soldering.

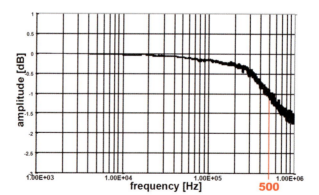

Fig. 12   IACS bandwidth: −3dB is above 500 kHz

The primary current bar is fabricated independently from the PCB, and subsequently glued to the backside of it.

Fig. 13   Three phase rectifier for high currents, and primary current bar concept

7   Conclusion

A new current sensor system, based on the AMR-effect, has been developed. Due to its external primary current bar concept, it can be adapted to a wide range of applications. Sensor chip, readout circuitry and biasing magnets are mounted inside a standard JEDEC SOIC16 package. First measurements show its targeted high dynamic range. The feasibility of this packaging concept was proven by a pre-qualification.

References

[1]   Loreit, U, Dettmann, F, Grundlagen magnetoresistiver Sensoren, 1992.

[2]   Schmitt, J, von Manteuffel, G, Hochdynamische Stromsensoren auf magnetoresis-tiver (MR) Basis, Sensor und Test Conference, 2010.

[3]   Loreit, U, Dettmann, F, Neuartiges magnetoresistives Feldgradienten-Sensor-Element und Stromsensorkonzept für magnetisch gestörte Umgebung, 1993.

[4]   http://www.circuitcalculator.com/wordpress/2006/01/31/pcb-trace-width-calculator/.

Andreas Nebeling, Wolfgang Schreiber-Prillwitz
Elmos Semiconductor AG
Heinrich-Hertz-Str. 1
44227 Dortmund
Germany
andreas.nebeling@elmos.eu
wolfgang.schreiber-prillwitz@elmos.eu

Keywords: current sensor, integrated sensor system, closed loop principle, system in package, AMR effect

Virtual Prototyping for Smart Systems for Electric, Safe and Networked Mobility

K. Einwich, Fraunhofer IIS/EAS Dresden

Abstract

The development of smart systems for mobility, especially the development of electronic systems for hybrid and electric cars, introduces a new dimension of complexity – the traditional separation of different physical domains and their corresponding engineering disciplines will be repealed. To address this new complexity new design methodologies are required. Due to several advantages virtual prototyping will become more and more important for the design of automotive systems. This paper introduces a modelling and simulation technology, which will help to overcome one of the major disadvantage of virtual prototyping, the long simulation time. Additionally this technology based on the description language SystemC/SystemC-AMS permits an IP protected model exchange through the automotive value chain from TIER2 to the OEM.

1   Introduction

The development of electronic components for automotive applications includes more and more the whole value chain from TIER2 semiconductor design houses via TIER1 system providers to the OEM car manufacturer. To realize new advanced features and to execute the new upcoming requirements for electric cars without exploding costs and to meet requirements like the energy efficiency, a tighter interaction of different design disciplines and physical domains will be required. Software becomes more and more involved, as it permits the production of highly configurable systems and an easy integration of sophisticated features. From the production cost point of view, software is the cheapest component, followed by digital hardware, followed by analogue electronics and components of non-electrical domains like mechanics. Thus there is a design trend in the opposite direction to replace partially the expensive components by cheaper ones. On this way e.g. mechanic or electronic components become simpler, however, to achieve the required performance they must be 'assisted' by digital hard- and software modules. The consequence of these design trends is that it is getting harder and harder – in some cases, even impossible – to consider the analog and non-electrical components of a system independently of the digital hard- and software ones or to design any IC without detailed knowledge of the full system environment.

Especially in automotive industry the overall system – finally the car – will be developed through cooperation of numerous specialized companies. The new design trends will make it more difficult to clearly specify interfaces and the behavior of components. Thus especially the concept and architecture design must be more and more done in a tight cooperation. To get the final concept and architecture numerous iterations between the partners are required.

For supporting the verification process inside those iterations several techniques are available. The most expensive way is the development of prototypes. Next from the cost perspective are Hardware in the Loop (HiL) simulations. Whereby the first introduces very long iteration cycles the second will permit only the simulation of components which can be simulated in real time (especially if there are analog and non-electrical components in the loop). Both techniques have limited introspection and debug possibilities and with both corner cases, tolerances and failures can hardly be checked. The third and usually cheapest variant is a purely virtual prototype and thus the simulation based verification of the overall system. Advantages are the very good introspection and debug possibilities and the possibility to simulate arbitrary scenarios, failures and to be able to introduce tolerances for system components and to check the behavior of an arbitrary constellation for those system variations. Disadvantage are the usually orders of magnitude higher simulation time, the effort which has to be spent for modeling, the verification and validation of the model and the risk that the model does not include all significant effects. Especially the very long simulation time very often makes the simulation of realistic scenarios impossible.

This paper will introduce a recent development for virtual prototyping which is especially applicable in the automotive domain. The discussed SystemC/SystemC-AMS based techniques addressing problems of virtual prototyping like the long simulation times. Also some for the automotive industry interesting feature of SystemC/SystemC-AMS like the possibility to exchange models in an IP protected way and to include them into arbitrary simulation environments will be discussed.

2   SystemC and SystemC-AMS

SystemC is a C++ based hardware description language with the focus on system architecture design for large digital hardware and software systems. It is hosted and standardized (IEEE-1666) [1] by the Open SystemC Initiative composed of many leading semiconductor companies and EDA vendors. Thanks to the nature of C++, SystemC is very flexible and powerful. In particular, SystemC supports methodologies that allow engineers to describe the interaction of hardware and software via highly abstracted modeling that facilitates access to very high-performance system-level simulations. To meet the requirements of heterogeneous systems, SystemC is being extended to the analogue domain. Also the SystemC-AMS extensions [2] are standardized and focus on abstract modeling to deliver overall system level simulations of real-time application scenarios, which require a simulation performance several orders higher than that achievable with models described in classical hardware description languages (see example in [3]).

SystemC / SystemC-AMS is a language based modeling approach. A sample SystemC description is shown in Figure 1. It is a simple so called primitive module which is a digital adder and a hierarchical module, which composes the adder and a latch to an integrator.

```
SC_MODULE(adder)
{
    sc_in<double>      in1;
    sc_in<double>      in2;

    sc_out<double>     outp;

    void do_add()
    {
        outp=in1+in2;
    }

    SC_CTOR(adder)
    {
        SC_METHOD(do_add);
        sensitive << in1 << in2;
    }
};
```

```
SC_MODULE(hier_arch_module)
{
    sc_in<double>      inp;
    sc_out<double>     outp;
    sc_in<bool>        clk;

    adder* i_add1;
    latch*  i_latch
    sc_signal<double> s1;

    SC_CTOR(adder)
    {
        i_add1=new adder("i_add1");
        i_add1->in1(inp);
        i_add1->in2(s1);
        i_add2->outp(out);

        i_dff=new dff("i_dff");
        i_dff->in(outp);
        i_dff->out(s1);
        i_dff->clk(clk);
    }};
```

Fig. 1.   Example for a SystemC primitive and hierarchical model description tion

On top of the language several vendors provide tools, which e.g. permit the composition of the models by a graphical interface (schematic editor). Using this graphical representation they automatically generate the language based description. Figure 2 shows an example of a graphical representation of the simple digital integrator example.

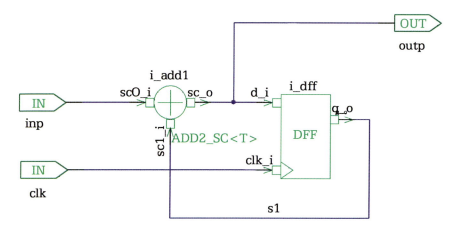

Fig. 2.   Example for graphical representation of a hierarchical description

The advantage of an approach based on a standardized language is, that models become exchangeable and composeable. That means independent from the concrete tool which was used to generate the model, the model will run in an arbitrary environment and also models generated by different tools can be connected. In this way models delivered by different vendors can be combined – a feature which is especially interesting for TIER1 automotive supplier. Thus today the most processor IP providers and manufacturers have SystemC models available for here customers. Compared to other hardware description languages, SystemC is not an independent language; furthermore it is based on the widespread programming language C++. This introduces a high flexibility, the usage of a lot of available libraries and an easy integration of software components. These features are especially useful for modeling at higher abstraction levels. Another specialty of SystemC/SystemC-AMS is the concept of Models of Computation. A hardware description language permits the description of a system in a way that the systems behavior can be calculated by a computer algorithm. Thus the algorithm, which is applicable to simulate a certain description, depends from the description themselves. SystemC/SystemC-AMS supports different modeling styles which permit the appliance of different algorithms, also called Model of Computation (MoC). The advantage of this approach is that in dependency of the problem and the used level of abstraction the most effective MoC can be applied. Due to such descriptions an optimal algorithm can be applied a very high simulation performance is achievable. In dependency of the concrete problem the possible speed up can be in the order of 1000 whereby in this case consequently behavioral modeling techniques has to be used [4]. This property makes SystemC/SystemC-AMS interesting for overall system functional modeling and thus the creation of a virtual platform which allows the appliance of real life application scenarios.

Whereby SystemC was developed for modeling large digital hard- and software systems, SystemC-AMS extends this approach to permit additional the abstract modeling of analogue mixed-signal behavior.

The standardized SystemC-AMS extensions support the modeling of non-conservative (directed signalflow) behavior using a synchronous dataflow and a linear signalflow MoC and conservative (networks) by a linear electrical network MoC. These MoC permit the creation of very fast virtual prototypes for a wide range of applications. However SystemC/SystemC-AMS is an open framework. Thus in [5] an extension was introduced which allows the description of conservative nonlinear dynamic systems. Figure 3 shows as an example the description of a dc-motor which uses this SystemC-AMS extension.

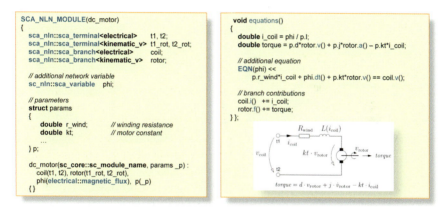

```
SCA_NLN_MODULE(dc_motor)
{
    sca_nln::sca_terminal<electrical>       t1, t2;
    sca_nln::sca_terminal<kinematic_v>  t1_rot, t2_rot;
    sca_nln::sca_branch<electrical>           coil;
    sca_nln::sca_branch<kinematic_v>    rotor;

    // additional network variable
    sc_nln::sca_variable    phi;

    // parameters
    struct params
    {
        double  r_wind;      // winding resistance
        double  kt;          // motor constant
        ...
    } p;

    dc_motor(sc_core::sc_module_name, params _p) :
        coil(t1, t2), rotor(t1_rot, t2_rot),
        phi(electrical::magnetic_flux), p(_p)
    {}
```

```
void equations()
{
    double i_coil = phi / p.l;
    double torque = p.d*rotor.v() + p.j*rotor.a() – p.kt*i_coil;

    // additional equation
    EQN(phi) <<
        p.r_wind*i_coil + phi.dt() + p.kt*rotor.v() == coil.v();

    // branch contributions
    coil.i()   += i_coil;
    rotor.f() += torque;
};
```

Fig. 3.   Example for the description of a primitive model of a dc-motor

3   Demonstration Example

As example for demonstrating the appliance of SystemC/SystemC-AMS in the automotive domain we used a window lifter. Parts of its model are shown in Figure 4. This model embraces components modeled at different abstraction levels and the use of different MoC. The module WINDOW LIFTER CONTROL contains a model of a microcontroller. There are two models of it at different levels of abstraction. Variant 1 is a bus cycle accurate behavioral model of the microcontroller, delivered by the controller IP provider, with the real software algorithms running on it. Variant 2 is a TLM based (Transaction Level Modeling, a special very efficient modeling technique [6]) algorithmic model. The variants can be selected by a parameter e.g. passed as a command line argument. Variant 1 will simulate slower than variant 2, however it will especially simulate the timing and processor load of the controller more accurate. For developing the algorithm and verifying the integration in the overall system the accuracy of variant 1 is usually not required. This accuracy is e.g. mandatory for checking that the processor finishes all tasks in time. However this can be done in a separate step after the algorithm development, where a longer simulation time is less critical. Thus the model speed and accuracy can be tuned to the model's use case. The digital-analogue, analogue-digital conversion and the analogue filtering is modeled within the dataflow MoC. The model of the non-linear mechanical subsystem shown in detail in Figure 4 uses a MoC based on nonlinear dynamic equations. Mainly, it consists of the motor, its driving circuit, the gear, the window pane and the limiters of the window frame.

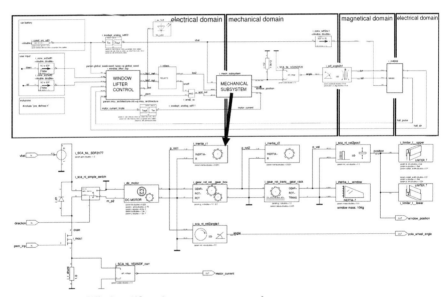

Fig. 4. Window lifter demonstrator example

Figure 5 shows some typical simulation results. The upper waveform illustrates the position of the upper edge of the window pane during the simulation. It can be seen that this upper edge hits an obstacle after about eight seconds. The software algorithms recognize it and the window pane is immediately lowered. After removing this obstacle, the window is closed until the pane reaches the upper limitation of the window frame. The waveform of the motor shaft's torque is shown in the middle section. Finally, the lower waveform shows the course of the corresponding motor current which is monitored by the software algorithms to detect obstacles. It can be observed that the current exceeds a threshold when the obstacle is hit.

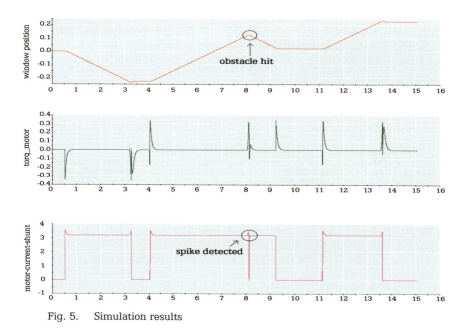

Fig. 5.   Simulation results

The simulation of this shown 15 seconds real time took less than a minute with the controller model of variant 2, using the cycle accurate controller model the same simulation takes about 3 hours. Thus this models permits the efficient development of algorithms and the verification, that tolerances does not lead to a system's failure (see also [7,8]).

4   Model Exchange

As mentioned before, one advantage of SystemC/SystemC-AMS is the possibility of an easy IP protected model exchange. Due SystemC/SystemC-AMS is based on C++ it will be compiled by a conventional C++ compiler. A compiled binary can only hardly be analyzed. In this way the model structure and the included algorithms will be protected. A customer who gets those pre-compiled model can analyze and debug parts of the model which the supplier has made available. Usually, this will be only the interfaces which are also available at the real hardware. To permit the development of embedded software special debug and analysis possibilities can be provided.

Most simulation, measurement and analysis environments provide a C-interface. These interfaces can be easily used to integrate a SystemC/SystemC-AMS model (C++ is a superset of C). Figure 6 shows the principle. The SystemC/

SystemC-AMS model will be embedded into a small wrapper library, which manages the time synchronization and data conversion. The model will be compiled and linked with this wrapper library and in dependency from the operating system a DLL or shared object will be generated. This dynamic library will be loaded by the customer tool or environment. From the customer perspective the SystemC/SystemC-AMS model looks like a conventional model of his tool or environment.

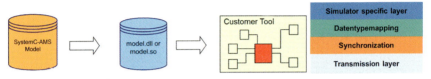

Fig. 6.   Principle way for model exchange and layered approach of the wrapper library

This model exchange technology can be principally applied to all tools and environments which provide a C- or similar interface. The wrapper library is based on the layered generic approach shown in figure 6. On this way it is easily adoptable by simply replacing the tool/simulator specific layer. Currently the library is available for tools like Mathworks Matlab/Simulink, VHDL/Verilog(-AMS) simulators like Spectre, NCSim, AdvanceMS, Modelsim and for the in the automotive domain widespread simulator Saber with the MAST modelling language.

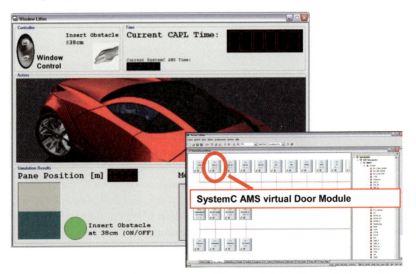

Fig. 7.   Integration of the Window lifter model in Vector Informatik's CANoe

Figure 7 shows as an example the integration of the described window lifter model in the tool CANoe from Vector Informatik [9] a tool for the development, testing and analysis of entire ECU networks and individual ECUs. In this demonstration a virtual module is integrated instead of the hardware module. This permits a start of the network development and analysis before the hardware availability. In this example a user can interact with the virtual door module via the CANoe graphical user interface similar like later with the real hardware door module due the achieved simulation performance permits a direct interaction.

5   Conclusion

The paper discussed a promising approach for virtual prototyping of complex electronic hard- software systems. It shows how the simulation time can be reduced to a level, that those virtual prototypes can be used to analyse the system's functionality. Thus the approach has the potential to replace partially costly hardware prototyping and HIL simulation. Additionally it enables new analysis and verification possibilities like for failure injection and the analysis of the influence of module tolerances.

Additionally the introduced technology permits the IP protected exchange of models of sub-components and their integration into arbitrary environments. This will enable a collaborative system development and specification through the automotive value chain.

6   Acknowledgement

The work was mainly executed in the context of the SUN project. This research project was supported by the German Government, Federal Ministry of Education and Research under the grant number 01M3178*. The author is responsible for the context of the paper.

References

[1]   IEEE Std 1666 - 2005 IEEE Standard SystemC Language Reference Manual IEEE Std 1666 2005 (2006): 0_1-423.

[2]   Open SystemC Initiative (OSCI) AMS Working Group: Standard SystemC AMS Extensions Language Reference Manual. 2010, http://www.systemc.org/downloads/standards/ams10.

[3]   Einwich, K., Application of SystemC/SystemC-AMS for the Specification of a Complex Wired Telecommunication System, 2005.

[4]   Viaud, E., F. Pecheux und A. Greiner. „An Efficient TLM/T Modeling and Simulation Environment Based on Conservative Parallel Discrete Event Principles." Design, Automation and Test in Europe, 2006. DATE ,06. Proceedings, 2006.

[5]   Uhle, T., Einwich, K., A SystemC AMS extension for the simulation of non-linear circuits, IEEE SoC Conference, 2010.

[6]   Ghenassia, F., Transaction Level Modeling with SystemC: TLM Concepts and Applications for Embedded Systems. Boston, MA, 2005.

[7]   Markwirth, T., Haase, J., Einwich, K., "Statistical modeling with SystemC-AMS for automotive systems." 2008.

[8]   Rafaila, M., et al. "Design of experiments for effective pre-silicon verification of automotive electronics." Specification & Design Languages, 2009.

[9]   Vector Informatik "ECU Development & Test with CANoe 7.5", http://www.vector.com/vi_canoe_en.

Karsten Einwich
Fraunhofer IIS/EAS Dresden
Zeunerstr. 38
01069 Dresden
Germany
karsten.einwich@eas.iis.fraunhofer.de

Keywords: virtual prototyping, modeling, simulation, multi-domain, mixed-signal, SystemC, SystemC-AMS, model exchange

Tool of Impedimetric Based Micro-Sensors for Research and Development of Electronic Components

A. Steinke, M. Hintz, S. Karmann, B. March, CiS Forschungsinstitut für Mikrosensorik und Photovoltaik GmbH

Abstract

Besides the use of sensors for applications in the automobile, there is an increasing demand for smart sensors that can be applied in the phase of research and development of electronic components for automobiles. For these requirements impedance-based micro-sensors were developed based on a technological platform. One example, the micro condensation sensor, is presented here. Thermally connected to the surface (PCB, components, connector), the microchips reproduce the temperature for the condensation process virtually without error. The detection system recognizes the condensation already in the range of droplet sizes smaller than 5µm and allows measurement up to sizes of 100 micron. The various applications of the tool require hybrid integration and thus guarantee a higher flexibility. There are calibrated, analog or digital output signals for temperature and condensed water mass available to the developer for appropriate applications.

1   Introduction

Recent trends in electronic assembling are leading to issues with regard to micro condensation - Examples include: minimizing the conductive track widths and spacing, increasing integration density, significant cost reduction, optimization of protection measures, use of the third dimension for wiring and for the assembly of the components, but also signal level trends with respect to field strength load, the use of different substrates in the assembly process and related use of new cleaning and protection materials. The results are breakdowns or failure accelerating effects caused by migration, corrosion and dendrites. [1, 2]

For cost and time reasons the design and technology development of electronic assemblies has to consider their use under the critical climatic conditions of micro-condensation from the beginning (moisture-temperature tests are not entirely relevant). Since "you can only test what you measure", the micro condensation sensors presented here are necessary.

The costs for protection of electronic circuit boards including housing are immense. Therefore, it is necessary to know the real micro climatic conditions occurring during field tests for all critical components, including the critical border area of condensation. In this paper miniaturized, artefact-free condensation sensors, are presented which represent the climatic conditions at different places in real time via an I2C bus. Results of field experiments in the car are shown.

The conditions determined in the automobile, in particular the condensation, have to be simulated in a climate test chamber. By means of the presented micro-condensation sensors it can be shown that during the condensation test according to ISO / DIS 16750-4 /3/ or other condensation tests there has been a stress for the assemblies. Results in different climate test chambers are presented.

2   The Danger of Condensation

The electrochemical migration in the form of dendrites is initiated by moisture in the form of condensates and ionic impurities. The dendrites are growing from the cathode to the anode. At the anode water is electrolysed and metal ions are dissolved. With an increasing level of impurities more metal ions will be dissolved. The metal-ions diffuse to the cathode via conductive path and deposit as metal there. [1, 4]

Fig.1.   Dendrite between anode and cathode

As a result of the higher conductivity between anode and cathode failure or malfunction (electric shorts) can occur. Therefore, it is very important to know: did or does a dew condensation take place, what quantity of dew condensation has accumulated, which temperature does the condensed water have and how long does the state of the dew condensation last?

To answer these questions micro-sensors are needed and one also has to measure the quantity of water condensation at different positions in a vehicle and to reproduce dew condensation at the same level in a climate test chamber. [5]

3   Micro Sensor System

3.1   Condensation Sensor System

For these applications a sensor with special features is required. Such sensor has to be miniaturized for reasons of assembly application on different places in the vehicle (electronic devices, plugs, printed circuit boards). The sensor must have a short response time and a low thermal mass. It has to operate with a low power consumption to ensure a minimum of thermal cross sensitivity and so as not to influence the thermodynamic equilibrium of condensation and evaporation. Further features are long term stability and high reproducibility of the measured parameters and chemical resistance to acid, alkalis and solvents for the completed micro system.

This means all components of a sensor system have to be considered already during the phase of development (filter, detection system, transducer, signal pre-processing, signal processing). In case of the developed micro condensation sensor, the detection system, transducer and signal pre-processing have been integrated in one package (see Fig. 2).

Fig. 2.   Micro condensation sensor

3.2 Transducer

The thermodynamic and electrical transducer has been realized in the construction of the silicon chip (Fig. 3).

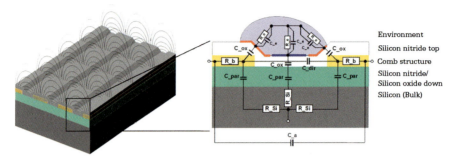

Fig. 3.   Detail of an interdigital structure with indication of the strayfield and diagram of the SPICE-model

Using the standard steps of an IC process the (integrated monolithic) temperature sensor is located close to the condensation area. If water vapour is condensed on the chip surface, the stray field capacity is changing significantly. The capacitor dimensions have been calculated by process and device simulation (FEM model). To approach a production stage the IC compatibility of the sensor process forms one of the most important issues in the development of process flows. Technology steps like photolithography, wet and dry etching, planarization, plasma enhanced chemical vapour deposition (PECVD, similar to passivation) and metallization do not affect the condensation area. The metallization layer consists of the typically used TiN and the passivation is a SiN layer realized by PECVD. This PECVD process also has to guarantee the reproducibility of the formation of water condensation nucleus.

3.3 Chip on Board (COB) Hybrid Integration

The design and the processes for embedding, bonding and encapsulation of silicon chips with an interdigital capacitor and an integrated temperature sensor require a very good thermal coupling of the detection system to the measuring object and a good thermal decoupling of the sensor chip from the signal processing. A COB-based hybrid assembly of the sensor components, detection system – transducer – signal preprocessing, places great demands on the impermeability between the glop-top and the sensor surface due to its application. In order to function, any creeping currents from the sensor surface to the electrical contacts have to be avoided. The application area of condensed water up to 90 ° C increases this requirement. Fig. 4. shows the mounting of the silicon chip in a FR4- cavity without and with thermal vias on the back side of the chip

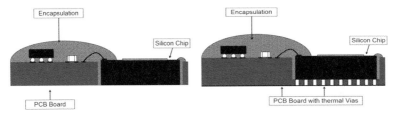

Fig. 4.   COB sensor device

3.4   Flip Chip Hybrid Integration

For the hybrid integration of sensor components specific requirements for the contact and conductive track system between the transducer and signal preprocessing are to be implemented. Therefore special technological micro-electronics compatible modules have been developed to meet these demands.

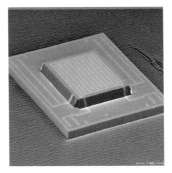

Fig. 5.   Sensor chip with lowered bond pads

For this purpose the bonding pads of the sensor chip are lowered significantly compared to the area of the stray field capacitor (200µm) (Fig. 5). With the flip chip assembly (Fig. 6) of this structure certain advantages can be achieved to the COB assembly. The contact systems are physically decoupled from the measuring medium. The technology is compatible for integration of signal pre-processing. There is a good thermal decoupling of signal processing and trans-ducer. A direct thermal coupling to the measuring environment is possible.

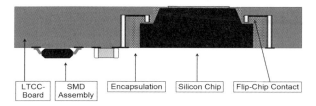

Fig. 6.   Flip Chip Sensor device

4   Signal Evaluation

Previously it was shown that the condensate composition can be determined optically [6]. The condensation was observed in a climate chamber and the water mass was determined. This method has the disadvantage visual access to the measuring site is needed in order to come to the corresponding results. Since the measurement of condensation happens in a thermodynamic range of high humidity, it can come to condensation may occur on the optical components which influence the result. Meanwhile, there are examples of a combination of optical and electrical condensation measurements (Fig. 7.).

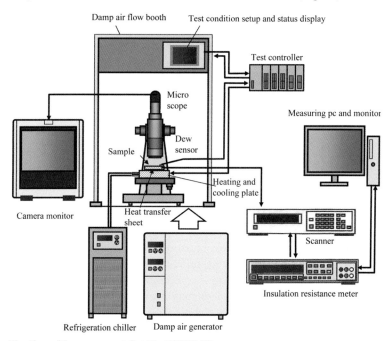

Fig. 7.   Measurement Set-Up ESPEC [5]

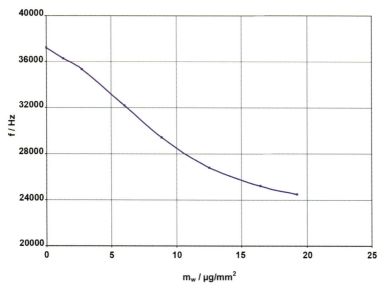

Fig. 8.   Non-calibrated output signal

Basis of the calibration of the micro-sensor systems is the determination of the water mass condensed on the sensor surface and its correlation with the sensor output signal (Fig. 8.). The water mass is determined by image processes and by knowing the contact angle of water droplets at the surface. A micro-controller translates the raw signal of the sensor to a calibrated output signal (0-1 V or I²C) (Fig. 9.).

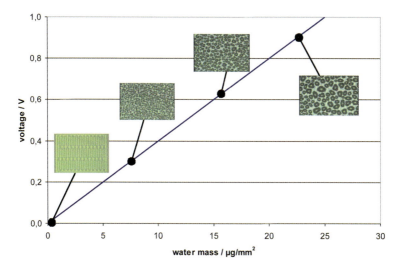

Fig. 9.   Calibrated output signal

5 Results and Discussion

5.1 Measurements in the Automobile

Current measurements in the car were made Fig. 10. shows a mounting place (HiFi-control unit). Measurements of the condensation at various points in the electronic system of an automobile were carried out repeatedly. Special emphasis was placed on recording the micro-climatic conditions both statically and dynamically. Important parameters for such a comprehensive description are the relative humidity and air temperature near the surface and the volume of printed circuit board assemblies, components, circuits, etc., their surface temperature and condensed water mass. From the air temperature and relative humidity the corresponding dew point temperature can be calculated and be taken in relation to the temperature of the surface. Considering the hydrophobic properties and pollution it can be concluded immediately on the appearance of condensation. From previous experimental series of car manufacturers partial condensation masses of 15 µg/mm² are known.

Fig. 10.   Mounting place in automobile [7]

5.2 Field Simulation in Climate Chamber

Condensation can not be detected visually in a climate chamber with electronic assemblies occupied at any point corresponding to the test cycle. However, the condensation sensors from the tool impedimetric sensors are well suited for that purpose. In this case it was shown that during the 3 h test cycle (Fig. 11.), condensation occured at a temperature of 25°C up to 80°C on the test probe (Fig. 12.).

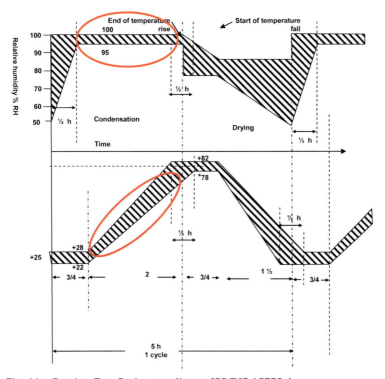

Fig. 11.   Dewing Test Cycle according to ISO/DIS 16750-4

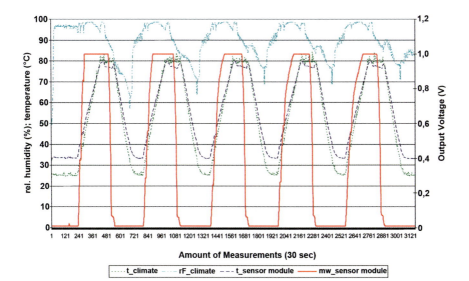

Fig. 12.   Measurement reading during Dewing Test Cycle

6 Further Applications

Besides the use of the basic sensor model and technology for the mentioned condensed water mass measurement, a special designed transducer (with a pitch of 3:4 of interdigital electrodes) and packaging has been integrated in a CCC* (condensate controlled capacitance according to HEINZE) dew point sensor. This sensor family can fill application niches such as long term atmospheric measurement in case of permanent high humidity (95-100% relative humidity) or in case of gas temperatures in the range of about 100°C and dew point temperatures up to 0°C. The use of such a sensor for inline field calibration of absolute humidity is another advantage.

The sensor platform is suitable for the use to optimizing the energy and material consumption in technology processes of electronic boards. The integration of such sensors in printed circuit boards during the phase of design and development provides early indication regarding degradation of materials or protection necessary to assure a high long-term quality.

7 Conclusion

In this paper a miniaturized impedimetric sensor has been described using the example of a sensor for measurement of condensated water mass. This sensor allows first time quantitative measurement of condensate mass with a calibrated output signal. The robustness of the micro-system solution allows the field measurement of condensation. We have presented the use of the sensor principle and different kinds of micro-mounting for other products such as dew point sensors or as test tool for the material and technology optimization of electronic boards, for example.

References

[1]   Schimpf, C., et. al., Failure of Electronic Devices due to Condensation, MicroNanoReliability 2007, Berlin 2007.

[2]   Scherl, R., Klimasicheres Design elektrischer und elektronischer Baugruppen; GfKORR- Jahrestagung "Beschichtungssysteme zum Schutz elektronischer Baugruppen, 2010.

[3]   DIN ISO/DIS 16750-4

[4]   Franke, J., Matzner, C., Einfluss von Schadgasen auf die Klimabeständigkeit elektronischer Baugruppen, 2. Landshuter Symposium Mikrosystemtechnik, 2010.

[5]   Tanaka, H., et. al., Investigate of Electrochemical Migration on Substrate Board under the Micro Dew Condensation Environment, Journal of Japan Institute of Electronics Packing; Vol. 13 No. 7, pp. 531-536, 2010.

[6]   Tanaka, H., Aoki, Y., Yamamoto, S., Evaluation Method for Ion Migration Using Dew Cycle Test (Part 3), ESPEC Technology Report No.4, 1997.

[7]   Picture courtesy BMW AG.

Arndt Steinke, Michael Hintz, Stephan Karmann, Barbara March
CiS Forschungsinstitut für Mikrosensorik und Photovoltaik GmbH
Konrad-Zuse-Straße 14
99099 Erfurt
Germany
asteinke@cismst.de
mhintz@cismst.de
skarmann@cismst.de
bmarch@cismst.de

Keywords: impedance micro sensor, condensation sensor, dewing test, dewing test simulation, electronic component reliability

Fraunhofer IMS is one of more than 50 institutes of the Fraunhofer Gesellschaft in Germany. With its competence in the areas of CMOS semiconductor components and technology, sensors and microsystems, circuit design and ASIC production the Fraunhofer IMS covers the complete potential of microelectronics. The Fraunhofer IMS has been certified according to DIN EN ISO 9001:2000 and the IC-Fab according to ISO TS 16949. The presented VGA-IRFPA is completely fabricated at Fraunhofer IMS on 8" CMOS wafers with an additional surface micromachining process for the microbolometers and the vacuum package.

2 IRFPA

Fraunhofer-IMS has developed an advanced digital IRFPA based on uncooled microbolometers with a pixel pitch of 25µm and a VGA resolution of 640 x 480 pixel. At a full frame frequency of 30Hz the IRFPA is designed for a high sensitivity (noise equivalent temperature difference NETD) of NETD < 100 mK. Due to the high dynamic range of a digital 16 bit output signal realized by $\Sigma\Delta$-ADCs a TEC-less operation in the temperature range between -40 °C and 80 °C is possible. The IRFPA needs two digital and one analogue power supply voltages. The technical data of the presented VGA-IRFPA are summarized in Tab. 1.

Parameter	Value
Image format	640 x 480 pixels
Frame frequency (progressive)	30 Hz
Output signal	16 bit (digital)
Temperature range	- 40 .. 80 °C
Power supply voltage	3.3 V and 1.8 V (digital)
	3.3 V (analog)
NETD	<100 mK (design value)

Tab. 1: Parameters of IMS VGA-IRFPAs

The block diagram of the presented digital VGA-IRFPA is shown in Fig. 1.

The 640 x 480 microbolometer array is read out by using massively parallel SD-ADCs located under the array. A certain amount of microbolometers is multiplexed by one SD-ADC. Blind microbolometers with a reduced responsivity are located in a ring arround the active microbolometer array. The blind microbolometers are read out identically to the active microbolometer. The SD-ADCs convert the scene dependent resistor change directly into 16 bit digital signals. These 16 bit image signals are fed into the digital video interface by a multiplexer. A sequencer controls the readout pattern by selecting each SD-ADC using a line and row control block. The configuration of the sequencer can be changed using an I2C-like interface. A built-in self-test supports the wafer test und reduces test time. The digital video interface provides three synchronization signals (horizontal, vertical and pixelclock) in addition to the image signals. A temperature sensor measures the temperature of the IRFPA. The temperature sensor is realized by three diodes connected in series followed by a buffer as an analog output stage. The temperature signal Vtemp is an analog voltage proportional to the temperature of the IRFPA and can be used for TEC-less operation. The IRFPA needs one analog supply voltage of Vdda = 3.3 V and two digital supply voltages of Vddd = 3.3 V and Vdd2 = 1.8 V. Five different reference voltages are necessary for biasing the SD-ADC. Apart from a reset signal and the I2C configuration signals the IRFPA needs only one digital clock signal [4]. The chip photo of the realized IRFPA is presented in Fig. 2

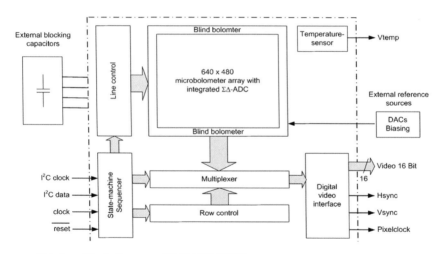

Fig. 1.   Block diagram of IMS VGA-IRFPA

The IRFPA is fabricated in a 0.35 µm CMOS technology with additional micromachining on top of the wafer at Fraunhofer IMS and occupies an area of approx. 326 mm2 with 13.6 million transistors. The chip-scale package has been removed for this chip photo. Most of the chip area is covered by the microbolometer array at the central part of the IRFPA. Two pad rows at the top and bottom edge of the IRFPA are integrated for the electrical connection of the IRFPA.

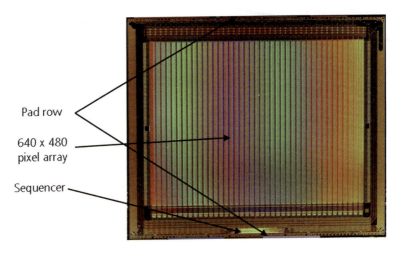

Pad row

640 x 480
pixel array

Sequencer

Fig. 2. Chip photo of IMS VGA-IRFPA

3 Bolometer

An amorphous silicon based microbolometer is employed as the IR sensor. The microbolometers are fabricated in a 25μm pixel pitch by post-processing CMOS wafers in the IMS Microsystems lab. The vertical and lateral pixel geometry was kept very simple and straightforward to ensure a solid baseline process with a high pixel operability [5].

A micromachined membrane is suspended approximately 2 μm above a metal reflector on top of the planarized CMOS and absorbs the IR-radiation. The resulting interferometric structure was numerically optimized for maximum absorption in the FIR band. The amorphous silicon sensing layer was optimized for low noise and high TCR. Two SEM microphotographs are shown in Fig. 3.

The left SEM micrograph in Fig. 3 shows a top view of a typical microbolom-eter. In the corners, the metal vias connecting the substrate to the sensing layer are located. Two insulating legs are defined along the edges of the central a-Si membrane. The right SEM image shows a cross section of these bolometers and demonstrates that low stress and therefore flat membranes have been achieved.

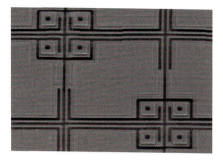

Fig. 3.   SEM micrograph of a bolometer (top view and cross section)

4   Vacuum Package

To reduce thermal losses by gas conduction a vacuum package with an infrared window is required. For reducing packaging costs Fraunhofer-IMS uses a chip-scaled package consisting of an IR-transparent window with an antireflection coating and a soldering frame for maintaining the vacuum. The realization and principle of a chip-scale package is shown in Fig. 4.

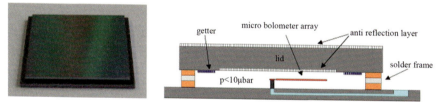

Fig. 4.   Realization (left) and principle of a chip scale package (right)

The IR-transparent lid consists of silicon with a double-sided antireflection coating and is placed using a solder frame on top of the substrate which includes the readout electronics and the bolometers. The use of silicon as a transparent lid results in lower production costs compared to germanium and causes lower mechanical stress due to equal expansion coefficients between the lid and the substrate. The chip scale package is currently under development in terms of long-time vacuum stability.

By using a flip-chip technique the lids are placed only on top of "good-tested" chips. This also reduces fabrication costs. Fig. 5 illustrates at the left a wafer with partly assembled chip-scale packages and at the right a chip-on-board mounting of this package onto a detector-board which is used in the IR-camera system.

The chip-on-board design has a size of 42 mm x 40 mm. The IRFPA can be connected from the back side of the PCB using a board connector.

Fig. 5. Wafer with chip scale packages (left) and PCB with IMS VGA-IRFPA (right)

5   Electro-optical Characterization

The IRFPAs are electro-optically characterized using a black body radiation source. The IRFPAs operated at an ambient temperature of 20 °C. The optical conditions correspond to f/1. A dedicated vacuum test package has been used during the characterization. The local distribution of the responsivity R is shown in Fig. 6.

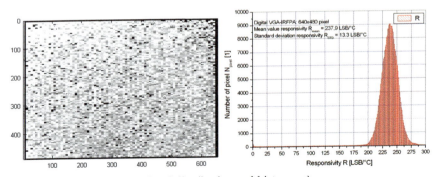

Fig. 6. Responsivity (local distribution and histogram)

The responsivity shown in Fig. 6 is calculated as the difference of the digital output values at black body temperatures of 25 °C respectively 35 °C. The local distribution at the left of Fig. 6 shows a homogeneous image. The right of Fig. 6 depicts the histogram of the responsivity which shows a Gaussian distribution with a mean value of R_{mean} = 238 LSB/K and a standard deviation of R_{std} = 13 LSB/K.

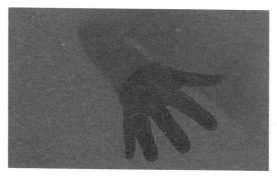

Fig. 7.   Uncompensated IR image with f/1.2 optics

An IR image of a human hand is illustrated in Fig. 7 using a f/1.2 lens. Apart from a simple offset correction the shown image is uncompensated, i.e. no gain, defect pixel, or noise correction has been applied. Further electro-optical characterizations are ongoing.

6   Conclusion

A digital IRFPA with 640 x 480 pixel with a 16 bit digital output signal has been designed, fabricated and electro-optically tested. The presented digital VGA-IRFPA is developed for pedestrian detection in automotive night vision systems. The microbolometers feature a pixel pitch of 25 µm and consist of amorphous silicon as the sensing layer. The digital readout of the microbolometer is based on a massively parallel use of SD modulators followed by sinc-filters. A chip-scaled package is realized for cost reasons as a vacuum package. For thermal imaging applications a chip-on-board solution is available.

Acknowledgements

The presented work is part of project "FIRKAM" funded by the German Bundesministerium für Bildung und Forschung (BMBF). The authors would like to thank the program manager "VDI Technologiezentrum" and the FIRKAM project partners for their helpful information and discussions. We highly appreciate the support of the engineers and technicians of Fraunhofer IMS during design, fabrication, and characterization of the IRFPA.

References

[1]   Reinhart, K., et al., Low-cost Approach for Far-Infrared Sensor Arrays for Hot-Spot Detection. In: G. Meyer et al., Automotive Night Vision Systems, Advanced Microsystems for Automotive Applications 2009, Springer Berlin, 2009.

[2]   Blackwell, R., Lacroix, et al., Uncooled VOx thermal imaging systems at BAE Systems, Proc. SPIE 6940, 694021 (2008).

[3]   Schimert, T., Brady, J., et al., Amorphous silicon based large format uncooled FPA microbolometer technology, Proc. SPIE 6940, 694023 (2008).

[4]   Weiler, D. Ruß, M., et al., A digital 25µm pixel-pitch uncooled amorphous silicon TEC-less VGA IRFPA with massive parallel Sigma-Delta-ADC readout, Proc. SPIE Conference Infrared Technology and Applications XXXVI, Volume 7660, 2010.

[5]   Ruß, M., Bauer, J., The geometric design of microbolometer elements for uncooled focal plane arrays, Proc. SPIE Conference Infrared Technology and Applications XXXIII, Volume 6542, 2007.

Dirk Weiler, Marco Ruß, Daniel Würfel, Renee Lerch, Pin Yang,
Jochen Bauer, Piotr Kropelnicki, Jennifer Heß, Holger Vogt
Fraunhofer Institute of Microelectronic Circuits and Systems (FhG-IMS)
Finkenstraße 61
47057 Duisburg
Germany
dirk.weiler@ims.fraunhofer.de
marco.russ@ims.fraunhofer.de
daniel.wuerfel@ims.fraunhofer.de
renee.lerch@ims.fraunhofer.de
pin.yang@ims.fraunhofer.de
jochen.bauer@ims.fraunhofer.de
piotr.kropelnicki@ims.fraunhofer.de
jennifer.hess@ims.fraunhofer.de
holger.vogt@ims.fraunhofer.de

Keywords: automotive night vision, far-infrared, bolometer, infrared sensor technology, sensor array, ROIC, vacuum package

Requirements of Processing Extended Floating Car Data in a Large Scale Environment

C. Oberauer, T. Stottan, R. Wagner, AUDIO MOBIL Elektronik GmbH

Abstract

Traffic volume is steadily increasing whilst infrastructure capacities stay the same – based on this prediction efficient processing of floating car data (FCD) and extended floating car data (xFCD) is of great significance to ensure future individual mobility. By taking into consideration results of existing FCD/xFCD studies the major subject of this paper is to define base requirements for collecting (x)FCD in a large scale environment. IM TD, AKTIV and COOPERS are some of the most significant projects in the field of car to infrastructure (C2X), car to car (C2C) and infrastructure to car (X2C) research sector. One focus of this paper is to analyse potential barriers of these approaches referring to a large area application with at least 10,000 xFCD data providers. As we see it an economically feasible solution for a road network wide capturing of floating car data is currently not available. The presented system requirements are the initial approach of processing xFCD in a large scale environment.

1 Introduction

1.1 Floating Car Data and Extended Floating Car Data

The idea of FCD is to collect real-time traffic data by locating the vehicle via GPS or GSM modules. Data such as car position, speed and direction of travel are generally understood as FCD [1]. Extended FCD cover a far wider set of vehicle data from different sources of the car. Depending on year and model the automotive CAN-Bus (controller area networking standard for automotive electronics) processes to a greater or lesser extent vehicle data. Ambient temperature, fog lights, windscreen wipers, switched-on lights, information from ABS and stability control systems are only a few examples of extended floating car data [2].

1.2 Problem Description

Floating Car Data (FCD) and Extended Floating Car Data (xFCD) are countlessly generated by moving vehicles. Extended FCD include a vast amount of data which already exists in millions of vehicles but there are only some research projects like SIM TD, AKTIV or COOPERS which collect and analyse xFCD of a limited number of vehicles and road segments. Extended FCD offer far greater possibilities beyond the detection of traffic situations but the huge potential which lies within hundreds of xFCD parameters cannot be exploited nowadays.

There are several barriers that have to be overcome before xFCD can be processed in a large scale environment. This analysis aims to summarize the difficulties arising from the above mentioned research projects and attempts to set base requirements for processing xFCD in a large scale testbed with 10,000 participants and above.

1.3 Related Projects

SIM TD, AKTIV and COOPERS are some of the most significant projects in the field of car to infrastructure (C2X), car to car (C2C) and infrastructure to car (X2C) research sector. The following consideration is not focused on evaluating achieved results of these well known projects, the main purpose is to analyse the potential barriers of these approaches within a large area application with at least 10,000 xFCD data providers.

SIM TD (Safe and Intelligent Mobility Testfield Germany) is set up as an operational test field for C2X communication. Various applications and services in the areas of road safety and traffic efficiency are/will be tested. The project comprises 400 sensor vehicles and 100 road side units (RSU). Beginning in 2008 and ending in 2011 the project is budgeted with approximately 69 Million Euros [3]. Breaking down overall costs to costs per vehicle results in 172,500 Euro per sensor vehicle. In case of SIM TD road side units are essential for wireless data transfer C2X and X2C. In context of implementing an xFCD system all over the road network the need of road side units is a major barrier for an extensive vehicle sensor network.

AKTIV (adaptive and cooperative technologies for intelligent traffic) is a German research initiative of automobile manufacturers and suppliers, electronics, telecommunication and software companies as well as research institutes. The main goals are to make traffic safer and more fluid with efficient C2C and C2X communication [4]. Beginning in 2006 until mid 2010 the project was budgeted with approximately 18.6 million Euros. On a road length of 15 kilometers in the area of Frankfurt the project consortium set up a working communication between specially equipped cars and intelligent road side infrastructure. Related to SIM-TD the need of road side units is a major barrier of a road network wide xFCD system, besides the incurred costs of 1.24 Million Euros per road kilometer on average.

COOPERS (Cooperative Systems for Intelligent Road Safety) is an EU funded research project which aimed to provide vehicles and drivers with real time traffic and safety related information on the high-level road network on various road sections all over Europe. An extensive enlargement seems to fail due to the high costs for necessary road side units. Approximately 16.8 million Euros were spent over a period of four years [5], but an economically feasible solution for a road network wide capturing of floating car data is not in sight.

2 System Requirements

By taking into consideration results of existing FCD/xFCD studies the major subject of this paper is to define base requirements for collecting xFCD in a large scale environment. Although the previous evaluation of current xFCD projects is strongly simplified it shows that there is still no satisfying approach for collecting xFCD within a large area available. As we see it xFCD system requirements should be divided in three sections: Basic system requirements, requirements for vehicle on-board-units and requirements concerning the xFCD processing software environment.

2.1 Basic xFCD System Requirements

▶ Vehicle data transmission over air interface with integrated GPS module and built-in SIM card (e.g. CDMA, GSM, UMTS and GPRS networks): An xFCD network covering the whole road network of an area has to be independent of cost-intensive road side units due to economic reasons. Communication over air interface is from today's point of view the most suitable approach in terms of costs.
▶ Vehicle platform independent on-board-unit: To be able to achieve a maximum possible penetration of xFCD sensor vehicles, the on-board-unit has to be suitable for all types and models of vehicles. This includes older vehicles and veterans by taking account to the fact that these can only send GPS data because of the missing CAN-Bus connectivity.
▶ Limitation of the network load with intelligent xFCD preprocessing in the vehicles: Intelligent preprocessing aims at limiting the network

load by processing collected data before transmitting. Depending on the required data sets the on-board-unit must be capable of processing constantly developing algorithms. This also can be realized by updates over air interface.

▶ Anonymity of xFCD data providers: This is a major criteria concerning acceptance of the xFCD system. Without guaranteed anonymity persuading car drivers will not be possible if a large amount of drivers is wanted.

2.2 Requirements for xFCD capable on-board-units

▶ CAN-Bus connectivity: Collecting xFCD of vehicles requires a CAN-Bus connection between the vehicle and the on-board-unit. Access to this critical vehicle area has to be read-only for safety reasons.

▶ Installation of the on-board-unit in all certified car workshops possible (max. 45 min. per unit): To ensure that the xFCD system is fully operational in all regions andcountries the installation process of the on-board-unit must be possible without special equipment and therefore practicable by every state-approved car workshop. A maximum of 45 minutes for each on-board-unit seems marketable.

▶ On-board-unit on-site maintenance free: Further requirement of handling a large number of on-board-units is that after the manual installation there should be no further garage appointment necessary. Due to most likely developments in the sector of data processing the on-board-unit has to be upgradeable with new algorithm software. This can only be realised with a cost-effective air interface.

▶ No energy consumption when the vehicle engine is turned off: To avoid a low or even empty car battery due to activities of the on-board-unit it has to be secured that there is no power consumption when the car engine is turned off.

2.3 Requirements for an xFCD processing software environment

▶ Platform independence of hardware and server software: Different regions and countries are used to work with different software platforms. In order to keep software adaption expenditure as low as possible the xFCD data centre has to be platform independent.

▶ Access point structure for providing individual data packets depending on the purpose:Incoming data is limited with preprocessing; outgoing data of the data centre has to be pooled depending on the purpose. Main objective of this data compression is to keep data traffic at the lowest possible level.

▶ Distribution of on-board-unit firmware updates: In addition of handling incoming and outgoing xFCD the software environment has to manage and distribute available software updates for the on-board-units.

3 The xFCD approach of Audio Mobil Elektronik GmbH

Traffic volume is steadily increasing whilst the infrastructure capacities stay the same - based on this prediction AUDIO MOBIL developed a real-time vehicle data management system (AMV®). Instead of offering yet another isolated solution, AUDIO MOBIL links existing systems and vehicles with AMV®.

Fig. 1.   On-board-Unit ASG®

The innovation is a powerful electronic component which transmits vehicle data over air interface to a data center where the data is processed and canalized. AMV® (Anonymous Management of Vehicle-real-time-information) is an extended floating car data management system, which has been derived from AUDIO MOBIL's research concept for traffic management. The idea is to create the basic conditions for the collection of Floating Car Data and extended Floating Car Data. Data like position and velocity (GPS based), but also CAN-based parameters like ambient temperature, pollution data, fuel consumption, fog lights or windscreen wipers can be read out. Each car, which has the on-board-unit ASG® (Anonymous Sensor data Gateway), functions as a mobile sensor, which delivers GPS and CAN-Bus data to the TrafficSoft®-data centre. This data centre is a data base, which manages the data for all subsequent purposes. A pillar of this concept is the consequent anonymity of data. Future aim is to generate a data carpet consisting of thousands of sensor vehicles, to be able to virtually cover the street with data.

3.1 Functional Overview on-board-Unit ASG®

▶ ASG® transmits existing real-time data from the vehicle. ASG extracts existing sensor data from the vehicle in real-time. This data can be preprocessed according to the intended use. The data is generated anonymously and is transmitted to the AMV-TrafficSoft® clearing house where the data is formatted for controlling and information purposes. (Registered patent FCD Gerät Patent number: 20 2010 004 383.2).

▶ Only the vehicle owner can annul the anonymity for specific sets of data. Every ASG® is assigned an ID that can be activated only by the vehicle owner. Thus the annulment of anonymity is solely the decision of the vehicle owner (e.g. for OEM services or cashless services like parking garages, road charges, etc.)

▶ Real-time data enhances the safety of the user. In case of theft or accident the vehicle is easy to locate but therefore an annulment of the anonymity is necessary by the car owner concerning the GPS signal.

▶ Cheap and easy to install in the vehicle. The small and flat module with integrated GSM and GPS antennas mounted behind the windscreen next to the rear mirror.

▶ Additional possibilities of connection / mountable without any hardware adjustments. The ASG® is connected to the chosen CAN bus and thus supplies the relevant vehicle data. By easily linking the ASG® to the CAN bus an adjustment of the existing hardware is not necessary. For more sensors more connection points are available. A novel energy management system makes the ASG® suitable for all vehicles. Energy efficient programming: no power consumption in the standby mode. (Registered patent Energiemanagement für Fahrzeuge Patent number: 20 2010 004 384.0). A wide range of input voltage opens up a flexible area of application – also for electric vehicles (9V–36V).

▶ Easy updating in the factory. New applications and data format changes are conducted by software updates via TrafficSoft®.

3.2 Functional Overview Software Environment TrafficSoft®

With TrafficSoft® AMV Networks GmbH provides a centralised, flexible and scalable software platform to receive, process and supply real time xFCD from any vehicle of the motorized individual traffic. Besides the management of real time vehicle sensor data TrafficSoft® is a flexible and scalable platform to integrate use case specific software modules which filter, process and supply the received data to the particular consumer. Furthermore TrafficSoft® is an important service component to manage the on board units within the vehicles (ASG®s) and keep them up to date. Additionally xFCD of third parties like OEMs or manufacturers of navigation solutions can be received and integrated into the data stream.

▶ Feartures of TrafficSoft® include: Anonymous management of vehicle real time sensor data: The software environment TrafficSoft® can process xFCD from ASG®s or other xFCD suppliers for instance OEMs or manufacturer of navigation systems. TrafficSoft® is capable of case specific delivery of xFCD –real time data of motorized individual traffic (no storage of raw data). It is safeguarding the anonymity of xFCD data providers à registered utility patent FCD-System (Patent number: 20 2010 004 382.4). It provides a protected web-portal for ASG® vehicle owners to locate the own vehicle and to check personal services. Finally TrafficSoft® provides a protected web-portal for certified ASG® installation garages for function check of assembled ASG®s.

▶ Management of the on-board-Unit ASG®: TrafficSoft® distributes automatic software updates over air interface to all ASG®s in the system. Updates can be necessary due to optimized CAN connection or use case specific special requirements. ASG® software updates provided via TrafficSoft® can be controlled selectively for specific vehicle platforms or even for single vehicles due to personal services. TrafficSoft® provides a vehicle manufacturer overlapping management system for CAN specifications[6]. This can be used for automatic improvement of the ASG® firmware and enables constant quality improvement of deliverable xFCD.

▶ Platform for individual application modules: It also serves as platform for use case specific software modules to process data individually. TrafficSoft® serves as platform for value added personalized services of external service providers and guarantees for safeguarding the complete anonymity of public utility data at the same time. TrafficSoft® offers the xFCD data providers the partial annulment of specific xFCD to take advantage of potential personal services in the future (e.g driver's logbook, mileage dependent insurance offers etc.). Finally it is capable of providing anonymous real time data streams for steadily expanding xFCD application areas.

4   Conclusions

The presented system requirements are the initial approach of processing xFCD in a large scale environment. Based on this concept AUDIO MOBIL Elektronik GmbH and AMV Networks GmbH developed a unique xFCD system which tries to unite findings of several xFCD research projects.

References

[1]   Leduc, G., Road Traffic Data: Collection Methods and Applications, European Commission, Institute of Prospective Technical Studies, pp. 5-7, 2008.

[2]   Busch, F., eSafety – Implementation Status Survey 2007 Technische Universität München, Chair of Traffic Engineering and Control, pp. 38-39, September 2008.

[3]   Barayou, K., Safe and Intelligent Mobility – Test Field Germany, Fraunhofer Institute for Secure Information Technology, (http://www.sit.fraunhofer.de/en/forschungsbereiche/projekte/simTD.jsp.)

[4]   Franz-Stöcker, U., Hessian Ministry of Economics, Transportation and State Development, Press Release, Sept. 2010.

[5]   http://www.coopers-ip.eu/index.php?id=36

[6]   CAN specification = vehicle manufacturer specific description of CAN data encoding of parameters

Christoph Oberauer, Thomas Stottan, Raimund Wagner
AUDIO MOBIL Elektronik GmbH
Audio Mobil Straße 5-7
5282 Ranshofen
Austria
christoph.oberauer@audio-mobil.com
thomas.stottan@audio-mobil.com
raimund.wagner@audio-mobil.com

Keywords: traffic management, FCD, xFCD

Appendices

Appendix A
List of Contributors

List Of Authors

Appendix B
List of Keywords

List of Keywords